The Physics
of
Manifestation

The Physics

of

Manifestation

Louis M. Houston

authorHOUSE®

AuthorHouse™
1663 Liberty Drive
Bloomington, IN 47403
www.authorhouse.com
Phone: 1-800-839-8640

Published by AuthorHouse   07/10/2012

ISBN: 978-1-4772-4056-4 (sc)
ISBN: 978-1-4772-4055-7 (hc)
ISBN: 978-1-4772-4036-6 (e)

Library of Congress Control Number: 2012912053

Table of Contents

Acknowledgments vii

Preface ix

Background 1

Properties of Thought 13

Manifestation 17

The Theory of Magic 25

The Theory of Probability Distortions 49

Concentration 67

Causality 71

Unconscious Concentration 73

Stability 77

Probability Waves 81

Power 83

Interference 87

Tuning 91

A Summary of the Theory 93

Acknowledgments

I am always grateful for the support of my family and friends. However, this book was largely inspired by an interest in explaining the traditionally inexplicable. I was heavily motivated by the book *The Secret* that made manifestation famous and demonstrated, through its overwhelming success the intrinsic desire that prods the individual to ignore skepticism in order to pursue their dreams.

Preface

The term magic has different connotations. It can be used to describe special moments when extraordinary things happen or it can more commonly refer to miraculous events which are inexplicable by science and are consistent with superstitious notions. The writer Arthur C. Clarke said that sufficiently advanced science can appear to be magic. That is, some phenomena occurs and we have no logical explanation for it, so we call it magic. We succinctly define magic as the process through which thoughts become reality. Consequently, if we marry a physics of thoughts to a physics of reality, we can produce a physics of magic.

That was the original goal of this book. There will always be a level of amazement when our thoughts become reality. We point out that although the normal manipulation of the body by the mind might be considered magical in this context, we prefer to focus on the ability of thoughts to manifest as remote physical phenomena.

An alternative term for magic as it is described here is manifestation. We find that manifestation is widely proclaimed but it retains a weak logical substructure in the manner in which it is presented. In general, vague terminology is used to describe manifestation, but it is strongly supported by its large interest.

In this book, we discuss manifestation and link it to a theory of magic. It is important to note that traditional magic is a broad and highly articulated system with a complex structure. However, we find that this structure is only weakly supported by scientific concepts and thus, we do not discuss the details of traditional magic in this book.

In this book, the physics of manifestation is developed with a background of a discussion of magic and it extends itself into a precise mathematical format. Manifestation is described as the result of probability waves which cause probability distortions that stimulate transitions of remote sites into desired physical configurations. Because manifestation is based on probability waves, there is no causal physical evidence that it exists. Rather, manifestation is supported by the coherent theory presented in this book and the statistical evidence provided by the successful manifestations which are experienced by many.

Background

The Buddha was born a prince and as such, was privileged to experience great wealth and privilege. In an effort to shield his son from hardship and suffering, the king constrained Buddha to remain on his palace estate until one day Buddha escaped his confines to explore the local community. To his great dismay and surprise, Buddha discovered old age, sickness, and death. This discovery led Buddha to leave his royal life, which included a wife and child, to search for a solution to the problems of life. In his six year sojourn, Buddha learned the mystic arts, including various forms of meditation and subjected himself to extreme asceticism. Finally, at the end of his ordeal, in a moment of clarity, Buddha realized that moderation was the key to salvation.

Buddhism began as a methodology for self-development approximately 2500 years ago and has since turned into a religion. The Buddha's main objective was the relief of suffering. Buddha determined that suffering is due to the act of clinging or attachment. Consequently, actions which lack attachment are promoted by Buddhism. The problem ensuing from attachment is the fact that everything changes and is thus impermanent. When the things we cling to change, we suffer.

Buddha's primary modus operandi for non-attachment is the middle path. That is, in the middle path we choose the central point between extremes. Following the middle path allows one to resist the attraction of extremes and to consequently exist in a state of non-attachment. Consistently, a famous Buddhist saying is that desire leads to suffering.

A major tool of Buddhism is meditation. Meditation occurs when one cultivates a heightened state of awareness of one's thoughts without posing any interference to them. Consequently, meditation promotes an unimposing awareness. As a result of meditation on a regular basis, one can help train the mind to avoid attachment.

Buddhism supports the idea that everything is connected through a form of interdependence. This type of connectivity gives rise to the notion of karma, which means action. Karma points to the causal connections between events and thus it allows one to connect the actions of a system of events to the consequences of those actions. We say that good karma is when good actions give rise to good consequences and bad karma is when bad actions give rise to bad consequences.

A fundamental idea of Buddhism is that at the core of reality is emptiness. Emptiness is the notion that a thing alone or in and of itself is empty, absent of identity and meaning. The reason for this emptiness is that an isolated system is without the benefit of comparison to other systems and without the relativity of comparison, there is no meaning. Consequently, for a thing in and of itself, existence is empty. The validity of the Buddhist notion of emptiness is often supported by the scientific fact that systems like atoms are primarily empty, consisting largely of space. As we will discuss later, emptiness has an even more important meaning than what is described here.

An essential philosophical concept which relates to Buddhism is the idea of dualism, the fundamental contrast of opposites and/or compliments. Almost all systems in reality can be divided into dual components. For example, hot contrasts with cold, male contrasts with female, good contrasts with bad, and human contrasts with god. When referring to dualism, we will use the terminology of extremes, as in one extreme exists in contrast to another extreme. Dualism is binary in that it consists of two components. Since Buddhism sees traditional human thinking as trapped in dualism, a major goal of Buddhism is to move beyond dualism to a state in which existential comparisons are unnecessary. This is the so-called enlightenment state, sometimes referred to as nirvana. Possibly the most noteworthy example which stems from dualism are the yin and the yang forces, products of the Chinese system known as the Tao. According to the Tao, the interaction between the yin and yang is responsible for all systems in reality. The yin is typically described as the more feminine and dark force, while the yang is typically described as the more male and bright force. The fact that negative is incorporated into the yin and positive is incorporated into the yang does not demean the male and feminine genders. In fact the male is often invested with yin while the female is often invested with yang. There is a level of homogeneity between yin and yang since there is always a little yang within yin and a little yin within yang. This homogeneity subtly projects the Tao beyond dualism.

The smallest complete system is tertiary or consisting of three parts. Examples include the three spatial dimensions, length, width, and height or the three spatial axes, x, y, and z. Other examples include the three colors red, blue, and green and the three states of matter, solid, liquid, and gas. We also have the three fundamental forces of nature, gravity, the strong force, and the electro-weak force and the three levels of polarity, negative, zero, and positive. A simple example of how being tertiary lends stability is a stool which is only stable when it consists of at least three legs. We want to focus on a particular tertiary system which represents structure. This system consists of an interior, a boundary, and an exterior. Consider that the unconscious mind is an interior, the conscious mind is the boundary, and the environment is the exterior. We can marry this tertiary system to the dualism of the inner universe and the outer universe, which we correlate to the unconscious mind and the environment. Think of a theme as a collection of related ideas or concepts. Ideally we would like the conscious mind to coordinate the relationship between inner themes and outer themes. That is, we want coherency between the inner and outer universe and this coherency is instigated from the boundary of the tertiary system. Themes behave like waves in the sense that they can interfere with one another. When two themes are coherent, they constructively interfere. When two themes are incoherent, they destructively interfere. We might think of constructive interference as a result of consistency and destructive interference as a result of inconsistency. Consistency generates positivity and inconsistency generates negativity. When the consciousness successfully correlates coherent themes in the inner and outer universes, positivity occurs. Alternatively, when the consciousness cannot produce coherent themes in the inner and outer universe, negativity occurs.

It is an interesting and pertinent point to make that complex systems are more subject to inconsistencies than simpler systems. The mind is composed of information sequences and information sequences are equivalent to energy. It follows that a mind consisting of more complex information sequences is more prone to inconsistencies and therefore it may sustain more negativity than a mind containing simpler information sequences. As a result of the balance between positivity and negativity, the energy in a consistent mind is greater than the energy in an inconsistent mind of comparable structure. Since energy is equivalent to intelligence, we can deduce that a consistent mind is more intelligent than an inconsistent mind of comparable structure. The mind is divided into the conscious and the unconscious compartments. Typically, the unconscious mind processes ninety percent of the total information in the mind. Consequently, a theme in the unconscious mind will be substantially more complex than the same theme in the conscious mind and as a result, the unconscious mind is more likely to contain inconsistencies than the conscious mind, even with respect to the same theme. In addition, if the theme itself is complex, the problem of inconsistencies will be significantly augmented, particularly within the unconscious mind. Generally speaking, themes are compartmentalized so that there is minimal interaction between distinct themes. This is consistent with reductionist thinking. In contrast, themes can become significantly connected, consistent with holistic thinking.

We introduce a concept called meta-duality which is the inclusion or superposition of extremes. Meta-duality moves beyond duality by choosing both extremes involved in a theme.

There are critical methods for developing meta-duality. One such method is the adoption of tertiary logic. The value of tertiary logic to meta-duality is the fact that it includes a natural interface between extremes. Tertiary logic includes three logical states: positive, negative, and zero or uncertainty. The major challenge is to identify uncertainty as a valid logical state. This is consistent with fuzzy logic. Fuzzy logic involves the concept of a membership function. When a logical system registers as belonging to a category which includes a range of values, the system is then a member of that category. In contrast, traditional logic, which includes true and false or positive and negative, respectively, would be constrained to register the logical system as either true or false. Comparatively, meta-duality would register a logical system as either equal to a specific extreme or belonging within the range of that extreme and its opposite.

Another method for developing meta-duality is the simultaneous practice of meditation and concentration. Meditation involves simple information sequences or low current, while concentration involves complex information sequences or high current. Meditation and concentration are the extremes of mental activity. Meditation trains the mind to ignore thought, while concentration trains the mind to control thought. Meditation is appropriate for the right brain, while concentration is appropriate for the left brain. The right brain is responsible for induction, philosophy and artistic thought, while the left brain is responsible for reduction, logic, and scientific thought. Concentration is similar to an engine, while meditation is similar to the coolant which constrains the energy dissipation of the engine. That is, when we meditate while concentrating on a complex information sequence, the energy dissipated by concentration is reduced. Consequently, the superposition of meditation and concentration produces a cooling right brain and a heating left brain which superpose a steady state energy dissipation.

Meta-duality benefits from the concept of a metaphorm. A metaphorm is an underlying superstructure which denotes the essential equivalence of apparently different concepts. An example of a metaphorm is the stimulation of flow which is represented by the concepts pressure and voltage. Pressure stimulates the flow of fluid, while voltage stimulates the flow of charge. Todd Siler originated the concept of metaphorms and used it to delineate the processes of the brain as they compare to cosmology. The term he used for his method is neurocosmology. In my book, *Metaphorms and Information Physics*, I presented metaphorms between the elements of electromagnetism and information theory. This included basic concepts like voltage, current, and resistance. These compare to meaning, information flow, and probability, respectively. It was found that information flow consistent with an information sequence is equivalent to energy. As a consequence, we assert that the total information flow in reality is conserved.

Buddha proposed that there is no such thing as a self or that the self is empty. That implies that a person has no intrinsic identity. We find that this proposal is fairly consistent with the nature of information produced by an information source. We might refer to an information source as a knowledge network. When a sequence is predictable, it contains no information and when the sequence is completely random, the information content is immeasurable. Consequently, a valid information sequence is only partially predictable. As such, the sequence contains no truly identifiable characteristics. This is consistent with the fact that all information contains some level of surprise. Since a person is identified by the information sequences they produce, it follows that there is truly no ground for identification. In this sense, the person doesn't have an identifiable self. However, it is more accurate to say that the self is fuzzy, because the associated information sequence remains somewhat predictable, and to that extent there is a somewhat identifiable personality.

This all suggests that if we maintain that a person has a highly defined self, that concept will be inconsistent with the truth. Alternatively, if we insist that there is no self, that concept will also be inconsistent with the truth. This is consistent with meta-duality as the superposition of extremes to the extent that meta-dual identification unpredictably alternates between the associated extremes. A meta-dual person is neither good nor bad and is also both good and bad.

We emphasize that at the outer level of reality with its many themes, concepts and representations, there is current flow and thus meaning. It is not until there is a thorough inner investigation of reality that one largely finds emptiness.

At this point, we need to discuss the nature of the information spectrum which characterizes thoughts. Thoughts contain different frequencies and those frequencies designate the nature of the thought. Symbolic thought is represented by language and occurs at the so-called visible range of information. The thoughts that we use in the everyday processes of life occur symbolically within our conscious mind. We "see" these thoughts and acknowledge them as largely representing our mental activity. However, thoughts occur in other frequency ranges. Emotions are thoughts with a higher frequency than symbolic thought. We are conscious of emotions as feelings though we don't represent them symbolically and emotions are generally not consistent with logic. Instinct comprises thoughts at frequencies below the symbolic or visible range. Instinct is very concrete thought, and as such is very connected to physical activity. Intuition is thought with a higher frequency range than emotion. At even higher frequencies are thoughts that we might classify as irrational. The unconscious mind processes thoughts at frequencies which are not visible to the consciousness. There is a continuous flow of thoughts in the mind which register at different frequencies as a multitude of themes are examined. Since we are only aware of symbolic thoughts or the thoughts associated with language, it is very difficult to coordinate all of our thoughts to behave in a coherent fashion. In order to create coherency throughout the mind, it is necessary to adopt themes which have integrity at all possible frequencies of thought. This is a challenge since it requires coherency at instinctive, symbolic, emotional, intuitive, and even irrational levels of thought. This brings to mind the concept of understanding. Normally, when we understand something, we develop a thorough symbolic description which generally does not include other thought frequencies. That is, when we understand, we develop a conscious awareness of a theme. A more deep-seated understanding occurs when all frequencies of thought are involved. In that case, both the conscious and unconscious portions of the mind become invested with the same theme. Meditation is an effective tool for expanding awareness from the conscious realm to the unconscious realm. We point out that the principle of non-attachment is a very powerful and positive tool for clearing the mind of confusing paradigms and addictive behavior, including lack of emotional control. Nonetheless, it is also through the adoption of meta-duality that a completely consistent understanding of reality ensues.

The difficulty associated with information processing is the abundance of conflicting and inconsistent themes within the outer universe. We live in an information age and information-based technology like computers and smart phones facilitate a diverse and eclectic lifestyle that keeps pace with the demands of society. Unfortunately, we are exposed to multitudes of themes which contain inconsistent structures. The conscious mind is compelled to sort through these themes and attempt to marry them to its internal thematic structure. A stable mental configuration exists when the mental structure has more integrity than the assortment of themes it encounters. Once meta-duality is adopted, the mind experiences a more stable structure, capable of enduring the influence of a full range of diverse themes. The real challenge is to condition the mind to behave in a meta-dual way.

Duality is a prime feature of reality. Consequently, people are heavily conditioned to utilize dualism in their everyday affairs. Unfortunately, dualism promotes the problem of attachment, because through dualism we are compelled to choose one extreme over the other. Nonetheless, dualism is a necessary and prevalent component of reality. For example, computer software is based on a binary language in which information is represented in terms of ones and zeros. Traditional logic derives from a truth table consisting of true and false. In addition, the basic understanding of a particular state is based on comparison to its dual partner, as for example, light versus dark. Let us not forget the fundamental dualism of mind and body. Obviously, because we are exposed to so much dualism, our minds become conditioned to develop themes with dualistic structures.

The challenge of meta-duality is to choose both extremes as equal partners without removing their distinctive values. In meta-duality there is a seeming dichotomy since we see the extremes as the same, but different. We refer to this experience as a stereo experience. It is easy to imagine how difficult it is to condition oneself to become meta-dual. For example, in meta-duality, we see no essential difference between pain and pleasure, though we retain the ability to identify them individually. Based on dualism, the typical person chooses pleasure over pain. There is no apparent motivation to choose both. However, it is proposed that by choosing the stereo experience of both extremes, meta-duality moves one beyond the experience of either extreme. At the point of movement beyond either extreme, identity becomes fuzzy and this state is synonymous with the zero-point energy within the void derived from quantum mechanics. The uncertainty principle in quantum mechanics designates the complimentary uncertainty between conjugate physical variables like position and momentum. When we know one variable well, we are compelled to have uncertainty about the other variable. This uncertainty leads to a zero-point energy within the ground state of every system, including the void. In this sense, meta-duality with the fuzzy superposition of both extremes, produces an intrinsic energy.

Meta-duality develops the mental state known as the omega (Ω) state. In the omega state, the mind is in a superposition of frequencies, rather than being in a specific frequency as in for example the alpha or beta states. Conditioning the mind to adopt the omega state largely consists of multi-tasking or multiplexing. In particular, the simultaneous use of the left and right brains is encouraged. The left brain tends to use active thought while the right brain tends to use passive thought. An active thought is consistent with speaking while a passive thought is consistent with listening. Alternatively, we might identify writing as active thought and reading as passive thought. When the mind is in the omega state, it is capable of sustaining an active and a passive thought simultaneously. The omega state is consistent with the superposition of extremes, and consequently, it is the ultimate objective of meta-duality.

Quantum mechanics is the study of the microscopic physical world. It fundamentally includes discrete energy levels, thus, the name quantum. The wave function describes the parameters of a quantum-mechanical system and characterizes the different possible states of the system in a superposition. When a quantum mechanical system is disturbed, it will collapse into one of the states represented in the superposition. We metaphorm a quantum-mechanical wave function into a meta-duality. The challenge is that we want our meta-duality to remain stable and not collapse when disturbed. This can happen if the meta-dual mind sustains a continuous train of thought that is not interrupted by exposure to other thought trains and disturbances. This becomes more realistic if the meta-dual thought train can be paused and held in stasis until its continuation. In that case, continuity exists only in the sense of a discrete information sequence. Essentially, the meta-dual mind is extremely focused.

It has been proposed that the omega state of the meta-dual mind is capable of sustaining a passive thought simultaneously with an active thought. We might designate this as the ability to identify two information sequences in an additive superposition. If we consider time multiplexing of the two sequences, the traditional process of recovering information from simultaneous sequences can be described as follows. There exists a sampling frequency based on regular clock intervals. During the first sub-interval a sample from the first sequence is acquired and during the second sub-interval a sample from the second sequence is acquired. This process repeats until the sequences are completely sampled. The original sequences can then be recovered by de-multiplexing the sampled data and interpolating the information. The ability of time-multiplexing to completely describe the information sequences is determined by the constraint that at least two samples are recorded at the highest frequency component of a given sequence. Consequently, the clock rate has to be sufficiently fast relative to the frequency content of the information sequences. We might, in this context, refer to an information sequence as either a current or a signal. The mental state which uses time multiplexing is constrained to have sufficient energy to accommodate the total energy of two simultaneous signals.

It is straightforward to imagine that meta-duality can incorporate the superposition of an arbitrary number of signals. In this case the signals represent more than the extremes of a particular theme. If we divide a theme into different tones or components, we might envision that meta-duality is able to incorporate the different tones simultaneously and distinctly recover each tone.

Because it simultaneously incorporates different themes, meta-duality designates uncertainty as the fundamental state of reality. From that uncertainty, energy is derived.

Properties of Thought

Given the full spectrum of thought which includes instinct, symbolism, emotion, intuition, and irrationality, at respectively higher and higher frequencies, we can envision a more complete form of meditation, which we shall refer to as high-frequency meditation. Normally, meditation is awareness and detachment from thoughts which are generally composed of symbolic information or language. High-frequency meditation includes the awareness and detachment from emotion and higher frequency thought. As a correlate to high-frequency meditation, there is also high-frequency concentration. Traditional concentration is directed at controlling thoughts. These thoughts are by default, symbolic. High-frequency concentration is directed at controlling high-frequency thought such as emotion.

Traditionally, meditation conditions the mind to relax and to detach from the influence of thoughts. However, in this sense, thoughts are in the form of symbolism or language. We modify this condition by broadening the definition of thought. Awareness of an emotional state is difficult to achieve when the emotion is subtle. It is even more difficult to control subtle emotions. We might extrapolate from high-frequency mental methods to broad-bandwidth mental methods. With such methods we gain the ability to either detach from or control thoughts of any frequency.

There is a dichotomy to the mind. Along with a high level of complexity and the ability to solve difficult problems, it comes with instability and a tendency to produce unpredictable information sequences. The solution to the undesirable symptoms is to imbue the mind with a system of coherent themes.

This requires a great deal of introspection and analysis. Caution is advised. While we may learn themes which solve problems, they may come with unwanted side effects. It follows that simple and robust themes are advisable like the golden rule or Occam's razor (i.e., the simplest solution is the best solution). However, we need a theme for different eventualities. In particular, subtle introspection requires subtle themes and such themes are generally complex. There is a natural paradox which occurs. In an effort to self-examine and simplify, we employ more and more complex themes. Such is the case for scientific progress. For example, unified field theories in physics are generally extremely complex. Ironically, their goal is a simplified picture of the physical universe. There is a question as to whether complex systems can be properly characterized by networks which are holistic or by networks which are compartmentalized. The answer to this question seems to depend upon the nature of the processor. In the brain there is an advantage to small localized subroutines as opposed to larger contiguous programs. An obvious advantage of this structure is trouble shooting or error correction. Another advantage of this structure is the ability to focus the processing effort into a single subroutine. Clearly, the only universal themes in such a processing structure must be fundamental. An example of a fundamental theme is mathematics. Another example of a fundamental theme is visual imagery. We offer an interesting combination of these themes in basic network theory. A network consists of nodes (N), edges (E), and faces (F). Euler discovered the equation $F-E+N-1=0$. We can use this equation to characterize networks. A processing system such as the brain consists of localized applications in which we can model Euler networks. Together these local networks interact to form the larger system which comprises the brain. In a dual system, the local network consists of the parameters $F=0$, $E=1$, $N=2$. We can interpret this as meaning that each substructure or subnetwork is characterized by these parameters. In a non-dual system, the local network consists of the parameters $F=0$, $E=0$, $N=1$. We point out that edges represent current, so that in the dual system, there is local current, while in the non-dual system, there is no local current. In a meta-dual system, the local network consists of the parameters F=1, E=1, N=1. In this case, there is local current, but only one node. We can designate the meta-dual network as a loop network because the edge forms a loop containing the single node. Most subroutines or substructures are dual, which means that the structure is divided into two complimentary or opposite sources (i.e., nodes). In contrast, a non-dual substructure consists of a single source. However, the traditional non-dual source produces no current since the network has no edges. It is only in the sense of the global structure that there are interacting sources. We can refer to the collection of time windows which de-multiplex into a processing substructure as a frame. A particular frame may consist of a multitude of interacting themes. During a small time period, the brain may be processing several frames at various frequency ranges. Noting the fact that voltage is proportional to current and that voltage is a metaphorm for meaning, we find that in a non-dual system, there is no local meaning. There is only global meaning. We might say that in a non-dual system, the frames are empty. In contrast, loop network frames contain meaning, although they are not dual. In other words, we might say that non-dual frames are gray, while loop frames are colorful. It is only the global ensemble of non-dual frames which present color. For example, Buddhism is a non-dual system and as such, it is meaningful only in a holistic sense. If we merge Buddhism with reductionist thinking, there is a loss of meaning because we are examining individual frames. Meta-dual systems or loop networks are compatible with reductionist thinking, because individual frames are not empty. It follows that dual systems are also compatible with reductionist thinking because of non-empty frames. Considering the possible configurations for networks, we compare the relative advantages. Typically, non-dual thinkers are only able to synthesize holistic theories. A meta-dual thinker can synthesize and reduce or use both global and local thought.

The existence of meaning is relative. Completely unpredictable or completely predictable information sequences have no meaning. Either we have to have a low enough intelligence so that normally predictable sequences are challenging or we have to have a high enough intelligence so that normally unpredictable sequences are tenable. An alternative view of intelligence is as a field which we absorb or radiate depending upon our intelligence potential. Intelligence is modified by our processing architecture. It is clear to see why intelligence may absorb meaning. We point out that inconsistencies may produce negative meaning. On the other hand, genius is challenged by what is normally extremely unpredictable. Such complex sequences would have no meaning to the average intelligence.

The single node in a loop network is a unimodal processor as opposed to the bimodal processor of the dual network. Unimodal refers to the superposition of holistic and reductionist processing. We have previously called unimodal meta-dual. In our original definition, meta-dual referred to a superposition of extremes. The idea is that the brain normally consists of a chain of bimodal processors, but an ideal state is for it to consist of a chain of unimodal processors or loop networks. This architecture minimizes the number of processors while maximizing the current. Consequently, the quantity of coherent information sequences is maximized.

Recall the concept of the frame, a system of coherent themes which is examined by the brain via time multiplexing. The brain multiplexes through many frames, but only one frame is visible at a given time. Frames can remain unconscious for long periods. Simple frames may consist of a single theme, while complex frames may consist of several connected themes. In a way, a frame is like a thesis or an in depth discussion.

Manifestation

Traditionally, the primary objective of life is the interpolation and extrapolation of frames. A frame may begin as intuition. It then becomes a feeling. Subsequently, it becomes symbolic. Next, it becomes instinct. Finally, it becomes a physical manifestation. Thus, the frame migrates from a subtle thought to physical reality. The previous designation of this migration or downshifting was tension leads to continuity. Tension correlates to thought and continuity correlates to manifestation, but frequency downshifting is more exact. The term intrapolation seems appropriate. Thus, we define intrapolation as the process of frequency downshifting of thoughts. Continued intrapolation will lead to physical manifestation.

Positivity reinforces itself while negativity diminishes itself. In other words, positive systems cooperate, while negative systems are disruptive. Consequently, at its core, reality is positive. We see this emulated in the atom, which has a positive core.

There is an essential uncertainty within reality. It is plausible to surmise that the tension caused by this uncertainty instigated a quantum mechanical disturbance in the void, initiating the universe. This explanation simply explains the Big Bang, the creation of the universe 13.7 billion years ago. The creative pressure of the essential uncertainty also explains evolution and its augmentation of consciousness.

Meta-duality seems dangerous, since we incorporate both the positive and the negative. This superposition is more safely accommodated as pure thought. Destructive interference is thus contained mentally until eventually, positivity gains the upper hand. The conditioning for meta-duality is thus a mental battle, until the equilibrium state is reached and positivity reigns. So, rather than meditate on the center point between extremes, we meditate on the interaction between extremes, giving each equal credibility. Meta-duality culminates into the omega state. The omega state is the result of an accelerated internal evolution. The omega state is free of attachment because it is the result of the tension of equivalent forces. Think of meta-duality as an interaction between two different frames, a positive frame and a negative frame. While the positive frame is eventually supported by cooperative themes, the negative frame is eventually dominated by one theme.

Meta-duality does not choose positive or negative. It is simply the nature of reality that positive dominates, eventually. It is not based on love or compassion, although these attributes are not inconsistent with positivity. Meta-duality is essentially based on strategy. Cooperation trumps single valued dominance.

Considering the frequency spectrum of thought, we can identify simple thought sequences as blue and complex thought sequences as red. We associate blue with cool and red with hot. Note that the relationship between energy and frequency for thoughts are inverse to that relationship for the electromagnetic field in the sense that lower-frequency thoughts are blue while high-frequency thoughts are red. Meditation and concentration are attributes of consciousness, the mind's awareness of thought sequences. When meditation occurs, consciousness shifts towards blue. When concentration occurs, consciousness shifts towards red. Consequently, concentration causes consciousness to dissipate more energy than meditation and the energy dissipation of consciousness is better controlled by the simultaneous use of meditation and concentration, which is consistent with meta-duality. Since meditation promotes the observation of thought, while concentration promotes the control of thought, meditation does not alter the frequency of thought, while concentration does. In fact, we find that continued concentration on a given thought sequence will simplify or blue-shift the thought sequence. This is consistent with the occurrence of manifestation in the sense that persistent concentration on a thought sequence will eventually shift the sequence into a physical manifestation of appropriate construction. Obviously, concentration is consistent with intrapolation.

We find that this result is also consistent with the nature of knowledge. Having knowledge of something gives one the ability to deconstruct it and reconstruct it. Essentially, knowledge of a thing objectifies it and separates it from the knower. When knowledge of a thought sequence is raised to a sufficient level, that thought sequence will manifest, as it develops a distinct, objectified existence in reality. We might say that manifestation occurs when we focus on a thought sequence to such an extent that the sequence develops a separate existence from our own psyche. Fundamentally, understanding delineates reality and in that sense, the manifestation of our ideas is a consequence of our thorough understanding of those ideas.

As manifestation is a consequence of concentration and since concentration is facilitated by the simultaneous use of meditation, which cools off the concentrating mind, we find that meta-duality facilitates manifestation. Meta-duality enables the mind to operate at a high level without burning out and high levels of concentration are necessary for successful manifestations.

The rather interesting thing about manifestation is that it points to the extreme connectivity of reality. This follows because in order for a thought sequence to finally manifest, it has to be transmitted at various frequencies through the network of channels created by the different interactive systems within reality. Furthermore, successful manifestation has to compensate for arbitrary disturbances of the thought signal as it propagates. A thought begins in the imagination. For purposes of consistency, we identify conceptual thought as imaginary thought. Consequently, we connect thought within the imagination to energy within reality. From this perspective, we see thoughts as existing within the imagination and manifestation as the process by which thoughts travel from the imagination to reality. It is clear that manifestation from the mind to the body is a continuous, natural process, but when we refer to manifestation, by default we are referring to transformations of thoughts from the mind to physical phenomena which are not directly connected to the body. Essentially, we are proposing that manifestation, in this context refers to the use of thought sequences to produce remote physical phenomena.

Manifestation is the process which transforms imagination into reality. However, there is a constraint. That is, what we create in our imagination must have the potential to be real. Reality unfolds in a stream of causes and effects. What happens today is a result of what happened yesterday. When we manifest, we insert a cause into the stream which is carefully crafted to yield a desired effect. It is clear that in order to manifest, we must have the sufficient freedom to alter the stream or what might have been our manifestation will be simply another outcome of pre-existing causes. In reality there is energy and that energy exists in different forms. When we transform imagination into reality, we convert a thought form into a physical form, but the energy is the same. The concept of emptiness is thoroughly related to manifestation because of the existence of forms and their essential equivalence. The essential equivalence of forms moves us beyond duality. Emptiness seems to mean lack of essence. A thing is made up of its parts and without those parts there is no thing. This is consistent with the notion of interdependence, however there is a subtle difference. Interdependence illustrates an object's meaning relative to other objects. Emptiness is a lack of essence because the object consists of parts and no individual part wholly represents the object. As those parts change, the object is transformed. In this way, the self has no essential existence. Energy has no essence because it has interchangeable forms. Basically, anything which has no specific form is empty of essence. Emptiness does not mean empty of energy or vibrancy. It just means that objects are composed of parts and without those parts there are no objects. I consist of a body and a mind, each of which is composed of parts. Without those parts, I would not exist. Because I consist of parts, I have no essence. Emptiness just means without essence. Everything is just a different form of energy. What does this mean for manifestation? Manifestation is the transformation of structure or form. Emptiness is form and form is emptiness. Things transform. They aren't created. Yet I do exist. What am I? I am a locus of interactions. There is certainly one substance: energy. We find that uncertainty is also empty because it has no delineation. It is that lack of delineation which produces the zero-point energy or the energy within the void. Because the energy has no specific identity, it is also empty. Frequency is form. I am a form of energy. My thoughts are a form of energy. When we manifest we transform thoughts into other forms of energy. First we transform higher frequency thought into lower frequency thought and then we transform that thought energy into probability waves. Those probability waves create distortions which transform probability. The new probability becomes energy and we have manifestation. Because of emptiness, we find that any object can be decomposed into parts. Thus, any thought can be decomposed into component thoughts. Those component thoughts can be further decomposed and so on. There are an infinite amount of different forms. However, energy is finite and the rate of flow from one form to another is limited by the conservation of energy. We distort probability and cause the manifestation of a form which is consistent with the structure of our thoughts. Positive and negative are just forms. The energy is essentially the same and there is no delineated essence. So, there is no real difference between positive and negative, other than form. This is where meta-duality, the simultaneous inclusion of extremes comes into play. There is no essential difference between pain and pleasure. There is no essential difference between a man and a woman. Compassion emerges from meta-duality because positivity is cooperative while negativity is not. All forms are impermanent. Information is a form, so it is not conserved. With all of this in mind, we can discuss the concept of enlightenment. It is just that like energy, enlightenment has no identity. Enlightenment is not a form. Enlightenment is energy. Since everything has energy, everything is enlightened. In order to realize our enlightenment, we must remove form. Energy is intelligence since energy is the flow of information. Enlightenment is intelligence with no structure. Thus, in order to realize enlightenment we must integrate all forms or become formless. Forms are consistent with duality. Therefore, in order to move beyond duality, integrate the mind and the body. Integrate the sciences and the arts. Integrate logic and emotions. Mathematics integrates all fields, so study mathematics. Essentially, we see that enlightenment is intelligence without a specific language or form of communication. The baby is enlightened until it learns to differentiate between positive and negative. The scholar is enlightened when she learns not to differentiate between positive and negative. Karma is a negative or positive reaction, based on the law of attraction or like attracts like. When we become enlightened, there is no karma, because there is no longer a difference between positive and negative.

Consciousness is form, so consciousness is not conserved. However, since intelligence is energy, intelligence is conserved. Furthermore, we find that existence is intelligence or energy and frequency or vibration is form. Manifestation is the process of giving form to energy. In that sense, manifestation is transformation. Concentration shifts frequency so it shifts form. Meditation interpenetrates forms and it is a way to connect to intelligence or energy. The universe consists of a maximal, but finite intelligence that is forever evolving. We might say that the universe is like a box with constant internal motion. Intelligence is the same with an infinite variety of forms. It is like the universe in a grain of sand. Intelligence is accessible. Intelligence is in the interrelationship of information. It is not the information itself. Rather, intelligence is manifest in the interdependence of forms. Energy is manifest in the interdependence of forms. Energy or intelligence is the one true thing which has no identity and thus no essence. An individual is a form of intelligence. Meditation and concentration are dual in the sense that meditation ignores form while concentration controls form. When intelligence remains formless, it is free. The wheel of life consists of the constant deconstruction and reconstruction of forms.

The problem which causes suffering is the attachment to forms. Since frequency is a form, higher intelligence is also a form. Intelligence is intrinsically universal. The enlightened are equally intelligent. Fundamentally, we all have the same intelligence potential. Truth is independent of form. Pleasure and pain are forms. When we are in pleasure, it is difficult to see the truth. When we are in pain, it is also difficult to see the truth. That is why an approach which neutralizes the effects of extremes is important. As long as we are locked in the cycles of pleasure and pain, we will never see the truth. The truth is that we are all energy or intelligence, and as such, we have nearly unlimited potential. But in order to realize our potential, we must learn to see beyond form and understand that everything and everyone are connected. When we elevate form, we produce pleasure. When we reduce form, we produce pain. When we remove form, we produce freedom. As a form gains energy, it elevates. As a form loses energy, it reduces. Manifestation is the process by which a form reduces mentally and elevates physically. Consequently, in manifestation there is a conservation of energy. In this sense, a form is a thought sequence. The extent to which a thought sequence is reduced in frequency is the extent to which a manifestation occurs. Forms appear to give identity, when in essence there is no real identity. One person typically sees another person as a form, allowing the one to identify the other. Identity is a property of duality, the existence of opposites or compliments. When we move beyond duality, we lose the notion of identity. Looking beyond duality, a person sees another as intelligence or energy and there is no distinction between subject and object. A cooperative interaction will disengage form, while a conflict will emphasize form. Consequently, positivity is consistent with the de-emphasis on form, while negativity is consistent with the emphasis on form. Nonetheless, positivity is elevated form. In the limit of infinite elevation of form, there is no form. This is the case of form of infinite frequency. The limit of evolution is the infinite elevation of form. This limit is never ultimately realized because there is only a finite amount of energy in the universe. However, freedom can also be realized by disengaging form. These are both cases of enlightenment. In the first case, enlightenment is achieved in an infinite period of time. In the second case, enlightenment is achieved in a finite period of time. Enlightenment consists of maximal intelligence and maximal freedom. Enlightenment is the realization of form-less energy. The energy within a closed system is conserved. The intelligence within a closed system is conserved. The total energy within the universe is zero. Therefore, the total intelligence within the universe is zero. In this sense, both energy and intelligence are without essence or empty. Nonetheless, the energy or intelligence within partitions of the universe remains substantial and conserved. A form is a closed system. When the mind/body is a form, it is a closed system and thus its energy is conserved. In that case, the energy cannot change. Nonetheless, all forms are impermanent. So, the mind/body changes and cannot truly conserve energy. The only way the mind/body can grow is when it relinquishes form and it is no longer a closed system. Meditation is a methodology for relinquishing form. A type of meditation is contemplation. Contemplation is a form of mindfulness.

The Theory of Magic

At this point, we want to define the process of magic as manifestation that occurs when thought sequences are used to produce remote physical phenomena. In this context, we find that manifestation is synonymous with magic.

We have already identified that successful information processing requires coherence between the inner and outer universe. In order to properly evaluate magic as a process, we must account for the inertia of existing concepts about the nature of magic. An accurate description of magic must minimally interfere with a priori descriptions. Nonetheless, false concepts of magic must be bypassed and replaced with more accurate and meaningful ideas. It has been stated that sufficiently sophisticated science is indistinguishable from magic. In this context, we propose that our analysis and propositions about magic are essentially science.

In our initial description, we propose that magic occurs in two general stages. The first stage is intrapolation. That is, the frequency of the thought sequence must be downshifted. The second stage is transmission. That is, the intrapolated thought sequence must be transmitted through space to its desired destination. Intrapolation is a product of concentration, so we will hence forth identify the first stage of magic as concentration. Recall that the thought spectrum range is instinct, language, emotion, intuition, and irrational thought. Therefore, concentration has the capacity to shift thoughts from the extreme of irrationality to the minimum of instinct. When a thought sequence is blue-shifted to the level of instinct, it develops physicality. At that point, the energy of the thought sequence is ready to be transmitted to its desired destination.

Let's first consider the stage of concentration. The essence of concentration is analysis of a thought sequence through logical reduction. We want to reduce the thought sequence into its fundamental components and then we want to produce synthetic models in order to reinterpret these sequences. Logical analysis includes descriptive language and symbolic manipulation such as scientific propositions and mathematical equations. We concentrate in order to thoroughly understand a thought sequence. Consequently, we are compelled to study the relationships within the sequence between the constitutive thoughts. There will also be relationships between the given thought sequence and external phenomena. In fact, the thought sequence can be in the form of a generalized proposition or law which has universal consequences. In contrast, the thought sequence may target a specific phenomenon. In either case, the goal of concentration on the thought sequence is to yield a precise description of the consequences of the manifestation of that sequence. The level of energy required by concentration depends on the frequency level of the thought sequence. Obviously, concentration on an irrational thought sequence dissipates the most energy. It follows that the most powerful magic is capable of manifesting irrational thought. However, this situation can often lead to chaotic and undesirable situations. The lowest energy concentration occurs for instinctive thought sequences. In fact, since the instinctive frequency is the lowest, concentration is not needed for this sequence. We will subsequently show that thought sequences which are originally instinctive are not projected by magical means. As mentioned earlier, the energetic demands of concentration are reduced by the addition of meditation. The previous comparison was that concentration is the engine and meditation is the coolant. Of course, this situation describes a meta-duality.

It is appropriate at this point to discuss meditation. By focusing on a simple repetitive pattern, meditation conditions the mind to become non-attached. Normally, meditation is the most effective when the body is in stasis, however meditation is primarily mental and consequently undiminished by bodily motion. A common form of meditation involves focusing on breathing, which has a regular pattern. I find that listening to music is meditative if the music has regular rhythms. Even walking can be meditative because it consists of a simple cyclic pattern. One of the objectives of meditation is to condition the mind to ignore thought sequences. Meditation is a property of consciousness and it brings consciousness into a cool, blue state. Since concentration heats consciousness, the superposition of meditation and concentration normalizes the temperature so that concentration can sustain a maximal level.

The brain consists of right and left hemispheres. For brevity, denote hemisphere as brain. The right brain focuses on inductive, philosophical and artistic thought while the left brain focuses on reductive, mathematical and scientific thought. In addition, the right brain is primarily emotional while the left brain is primarily logical. Consequently, the left brain is the best at concentration. We propose that meta-dual functioning of the right and left brains requires that the right brain meditates while the left brain concentrates. We can interpret this as the superposition of passive thinking in the right brain with active thinking in the left brain. In this configuration, the right brain cools while the left brain heats. The overall brain "temperature" is moderated by this superposition.

The second stage of manifestation is the transmission of energy to its remote destination. This stage requires visualization. We need to visualize where we want the energy to be transmitted. The process of visualization requires imagery expressed by graphics and/or language. Visualization is a process best performed by the right brain. While concentration correlates more with theory, visualization correlates more with experiment. Consequently, concentration and visualization complement each other in a similar manner to the way theory and experiment compliment one another. At this point, magic incorporates the two stages of concentration and visualization enacted by the left and right brain, respectively. Concentration intrapolates the original thought sequence down to the physical level and then visualization teleports the energy of the thought sequence to its desired destination. Visualization is enhanced by diagrams, chants and icons.

Traditional magic is a highly complex system with many elements. We won't even attempt to describe this system here, but we can identify the most popular magical element as a spell. The spell is the vehicle through which magic is primarily cast. We commonly hear the statement that the magic user casts a spell. While there is no sufficient evidence that the system of magic is valid, due to the inertia of common knowledge and in the interest of coherency, it is important to validate the concept of a spell. We have already described that valid magic consists of concentration and visualization. Concentration creates a low frequency thought sequence which is cool and highly consolidated. By consolidation we mean that the understanding of the thought sequence is high enough to make the sequence instinctual to the magic user. We will refer to this as a consolidated thought sequence. After consolidation, the sequence is visualized at its final destination. We propose that the visualization process correlates to a spell. It's clear that visualization must encode the consolidated thought sequence. From this perspective, we can interpret a spell as a code. Obviously, the code has to be interfaced with the consolidated thought sequence. We can argue that a code is a visualization tool since it is a unique signature. In this venue, a code represents a unique pattern for a specific point in space and/or time. As an example of a magical system, consider the original thought sequence being the desire to communicate over a remote distance. The first stage is to concentrate on that thought sequence. After a certain period of concentration, the reality of this thought sequence becomes a telephone. The next stage is to visualize a destination for the thought energy and subsequently transmit the signal. This requires a code or a spell. The code is thus a telephone number which uniquely designates a specific destination for the energy signal.

Consider another example of magic. Einstein first imagined what would happen if he could ride a light beam. This is an intuitive thought sequence. After significant concentration, the thought sequence became consolidated as the special theory of relativity. It is at the experimental stage that the theory is fully consolidated. Manifestation of this theory was realized when time dilation effects on microscopic particles were measured. This theory was encoded for the education of the public by a set of specific equations which specify the effects of motion on fundamental quantities like mass, length and time. In the way that Einstein's story correlates to magic, it is clear that the equations of the theory represent the code or spell. Using this example, we can further delineate the definition of a spell. A spell is an arrangement of symbols whose connectivity designates specific relationships or rules. A spell encodes the fundamental elements of the associated consolidated thought sequence and it implicitly specifies its destination.

Magic must begin at a high frequency, higher than instinctive thought. This requirement is based on the equation which says that the imaginary thought energy equals the energy of consolidation plus the energy of transmission. When the imaginary thought energy equals the energy of consolidation, the thought is not imaginary but real and there is no residual transmission energy. The energy of transmission is normally loss when thought energy becomes real. Magic conserves this energy and consequently makes the conversion from imagination to reality completely efficient. A thought sequence that is initiated as instinct cannot be used to perform magic, because there is no transmission energy. Alternatively, the higher the frequency of the original thought sequence, the more effective the transmission of the sequence. For example, a thought sequence that begins as intuition will manifest more strongly than a thought sequence that begins as emotion and a thought sequence that begins as emotion will manifest more strongly than a thought sequence that begins as symbolic language. However, magic always exacts a cost which largely depends on the level of concentration. Intuitive thought sequences require more concentration than emotional thought sequences, for example. We shall refer to a spell which transmits an originally symbolic thought sequence as an X spell. For originally emotional thought sequences, we use a Y spell, and for originally intuitive thought sequences, we use a Z spell. Obviously, Z>Y>X. If the thought sequence is originally irrational, we use a Q spell. A Q spell is not necessarily subject to physical laws. It is obviously more powerful than a Z spell.

It is interesting to observe that since animals largely use instinctive thought, they are largely incapable of magic. We can summarize that magic consists of concentration and a code, which we identify as a spell. The concentration consolidates high frequency thought and the spell symbolically represents that thought as it can be interpreted in a remote location. It is essential that the destination of a spell has a compatible matrix for the thought sequence being transmitted. For example, one individual may cast a spell that is transmitted to other similar individuals. For the sake of brevity, we shall refer to a matrix as the destination of a spell. In that sense a matrix can represent a range of real systems and it can be of any size. Thus, magic requires concentration, a spell and a matrix.

Observe that magic cannot be produced solely through meditation. In fact, meditation is only useful as a coolant for concentration. While the concentration stage of magic is largely left brained, the spell stage is largely right brained. This is true to the extent that a spell incorporates symbolism and synthesis.

Concentration produces intrapolation, leading to consolidation of the original thought sequence. Without intrapolation, there is no residual energy for transmission of the spell. Normally, a magic user must originate their own spells. It is a fallacy to believe that spells can be cast by anyone, because if we are presented with a consolidated thought sequence and we wish to transmit it with a spell we need the energy required to transmit the sequence. That energy is only available through either intrapolation or extrapolation. When we extrapolate the consolidated sequence, we migrate it to its original frequency, which must be a higher frequency. Extrapolation also requires concentration. When we concentrate in this way the thought sequence gains energy and that energy is of the specific nature required for that sequence to be transmitted to the matrix by the appropriate spell. As an example of extrapolation, suppose that we are presented with a set of equations that represent a spell. In order to cast that spell, we must understand those equations to a sufficient level that allows us to know the original theory which generates the equations. As another example, consider a more traditional spell which is in the form of an incantation. In order to successfully cast that spell, we must be able to understand the original theory which generates the incantation. In both cases, concentration is required. Thus, we once again conclude that magic includes concentration, a spell, and a matrix.

At this point, we must discuss the example of the telephone. As magic, it includes the telephone device as the product of concentration, the phone number as the spell, and a receiving telephone as the matrix. It is clear that an arbitrary individual does not understand the intricate machinery of a telephone, but that individual can readily use a telephone. This seems to present a contradiction. The fact is that using a telephone is not magic. It was only the first use of a telephone that represented magic. The same applies to Einstein's theory of special relativity, as it does to most of science. Magic is only present at the initial use of invention. Magic introduces a new logical substructure into reality. In traditional magic, this substructure is temporary. In science, this substructure is permanent. The initiation of a logical substructure is equivalent to a probability distortion. That is, either a normally improbable event becomes probable or a normally probable event becomes improbable. For cases in which magic creates science, the probability distortion becomes permanent. As discussed in the book, *Metaphorms and Information Physics*, intelligence is capable of producing probability distortions. This is consistent with the fact that magic requires concentration and visualization, which are products of intelligence. Essentially, in the process of manifestation, magic produces probability distortions with concentration, a spell and a matrix. When the probability distortion is permanent, magic becomes science.

A matrix is a network which is activated by the absorption of energy. When we think of a human being as a matrix, we can envision the activation of physicality or mentality as the result of energy absorption. When we think of a machine as a matrix, we can envision the motion of components as the result of energy absorption. A matrix varies in size and complexity. For example, a matrix may be a single individual or it may be a community. It may be a bicycle or it may be a computer. The size of a matrix used in magic relates to the intensity of the energy signal. The intensity of the signal varies inversely as the square of the distance between the source of the magic and the matrix. Consequently, the distance between the spell and the matrix determines the size of the matrix. For a given spell, if the matrix is close to the spell, the matrix can be large. If the matrix is far from the spell, the matrix will tend to be small. The size of the matrix also depends on the specificity of the spell. If the spell is specific, the matrix will be small. If the spell is general, the matrix will be large. In summary, the size of a matrix depends on the intensity of a spell and on the specificity of a spell.

We want to consider a case when concentration does not consist of logical analysis. Identify a sorcerer as the practitioner of magic. Rather than traditional concentration, suppose that the sorcerer engages in belief. That is, the sorcerer, based on some evidence, simply believes in a thought sequence. Depending on the thought sequence, the ability to believe can be very difficult. Belief manifests when the believer is more consistent than the matrix. Energy flows from consistency to inconsistency. This explains why Santa Claus remains a myth even though many children believe in him. Those believers would have to be more consistent than the outer universe in order for Santa Claus to manifest. This explains why Christ was able to perform miracles, because he was more consistent than physical reality. The only requirement for manifestation was Christ's belief. It follows that the level of manifestation depends on the level of consistency of the believer. It is said that when people of faith believe in a phenomenon, that phenomenon happens. The extent to which that phenomenon happens depends on the strength of the belief and the level of consistency of the believer. The ability to use belief as concentration is the ultimate form of magic. As an alternative example, an insane person may have many irrational beliefs, but those beliefs are not grounded because of the many internal inconsistencies of the insane. As a consequence, those beliefs don't manifest in reality. This brings to mind the challenges of manifesting irrational thought. In order for this to happen, the thought must envelop some consistent structure that is somehow coherent while simultaneously being irrational.

The issues related to belief have several consequences. The inability to manifest with belief is due to inconsistencies in the inner universe. The minds of most thinkers are compartmentalized so that positivity is isolated from negativity. This is the leading cause of emotional mood swings. Inconsistencies represent negativity. Consequently, it is the presence of a net negativity in the inner universe which disables the ability to manifest with belief. Consider a child which exists in an innocent state. Nonetheless, the typical child compartmentalizes positivity and negativity. This is evidenced by the child's tendency to undergo emotional cycles. A child believes in fairy tales, yet those tales remain fictional because the child is insufficiently consistent. We can conclude from this that the compartmentalization of positivity and negativity is a natural state. The development of holistic thinking so that positivity and negativity are combined is the goal of meta-duality. Meta-duality creates a synergistic interaction between positivity and negativity so that coherence is reached. The meta-dual sorcerer is capable of manifesting with belief because his inner universe is consistent. When concentration becomes belief, there is no spell. Consequently, there is no transmission energy due to a downshift in frequency. Transmission naturally occurs because of the rule that consistency sends energy to inconsistency. In other words, the consistency of the sorcerer is the driving force that transmits the sorcerer's belief to the matrix.

There is an observation to be made about consistency sending energy to inconsistency. As we grow and learn as individuals, our studies and experiences in the outer universe typically cause us to become more consistent with the outer universe. However, our maximal level of consistency is limited by the consistency of the outer universe. It is only through internal focus and individual study that we can elevate the consistency of our inner universe to a level beyond that of the outer. The sorcerer understands this and thus, extends his study of the universe to a personal perspective. Once you have learned all you can learn from the outside, you must then look inside and learn more.

Manifestation is very popular. It was mainly popularized by a book called *The Secret*, in which the law of attraction was discussed. The law of attraction essentially says that like attracts like. That is, if we think good things, then good things happen or if we think bad things, then bad things happen. *The Secret* promotes the idea that if you focus on positive things, then they will be attracted into your life. From the perspective of most who endorse *The Secret*, this is all consistent with Y-spells, in which the original thought sequence is emotional. A Y-spell is driven by some type of desire. Examples of success using the law of attraction are people who have found fame and/or fortune. In these types of scenarios, people are advised to act as if their desire has already been fulfilled. That is, if your desire is to be rich, then do the things that rich people do. This is consistent with the law tension creates continuity in the sense that acting out the fantasy produces the tension and reality creates the continuity by turning the fantasy into reality. Acting out the fantasy is a spell, since it is a graphic visualization of the original thought sequence. We can easily see that spell is to tension as manifestation is to continuity. There is also concentration involved in this. Concentration produces the script that we act from. If we are pretending to be rich, we need to decide what types of actions the rich participate in that do not necessarily require wealth so that we may emulate them. In general, whatever our desire is, we have to figure out how to act in such a way that our desire appears to have been fulfilled without actually having been fulfilled. This is a form of problem solving which necessitates concentration.

There is a motive force underlying each thought frequency. High frequency thought naturally tends to reduce to low frequencies. We have already pointed out that the motive force for emotional thought is desire. Symbolic thought sequences are motivated by the property of information that it will not remain isolated. That is, communication is a property of language. Wittgenstein said that there is no such thing as an isolated language. Thus, even without desire, people will tend to communicate. Intuitive thought sequences represent knowledge. Knowledge resolves itself. When Einstein imagined riding a light beam, he was having an intuition. When his theory of special relativity emerged, his intuition had finally resolved itself. Irrational thought sequences are indicative of disorder. Disorder spreads and grows. Hitler's pogrom against the Jews might be considered a manifestation of irrational thought. Since these thought frequencies are driven by their own motivations, why is concentration needed? Concentration is required in order to make the original thought sequence understandable. For example, even when a person is compelled to communicate, she must take the time to organize her thoughts so that her words are understandable.

Consistent with our current description, let us review the three components of magic. They are concentration, a spell, and a matrix, unless the concentration is replaced by belief, in which case there is no spell needed. The concentration is some form of logical analysis, the spell is either a description, a set of equations or a graphic illustration. The graphics may be auditory as in some form of music. The matrix is the system that receives the magic from the spell. The distance between the spell and the matrix partially determines how intense the magic is. The magic's intensity is also dependent on the frequency of the original thought sequence. There are four types of spells, an X-spell for symbolic frequencies, a Y-spell for emotional frequencies, a Z-spell for intuitive frequencies, and a Q-spell for irrational frequencies. X-spells are motivated by communication, Y-spells are motivated by desire, Z-spells are motivated by knowledge, and Q-spells are motivated by disorder.

There is something to be said about the channel which transmits energy from the spell to the matrix. In traditional magic, this channel is not generally categorized since magic apparently transmits instantaneously across time and space. However, since in this theory of magic, the spell is real, it is constrained to send its energy through space in a causal fashion. The energy of the spell can be of any physical form, electromagnetic, acoustic, heat, mechanical, and so forth, but the most important element of transmission is the incorporation of the energy into the logical substructure of the matrix. In this sense, the spell connects to the causal chains which link the matrix to the environment. That environment contains the channel which connects the matrix to the spell and exploits the causal interconnectivity affected by the logical substructure of the spell. This is very important because it demonstrates that a spell can create a specific disturbance that will causally propagate to the matrix. This can happen in a way which renders the transmitted energy as virtually undetectable. For example, consider the butterfly effect: a small disturbance at one point in space can cause a large disturbance far away from the original disturbance. The manipulation of these subtle causal chains can occur via small seemingly random actions of the sorcerer. It is the spell which influences the sorcerer to initiate causal disturbances. Those disturbances may in fact be more significant than the energy of the spell in propagating towards the matrix. In fact, silent, localized spells can still be quite effective because of the induced behavior in the sorcerer and the radiation of that behavior into the logical substructure of the matrix.

We find that although it may not be the only means, meta-duality promotes internal consistency and it is a high level of internal consistency which enables the sorcerer to cast magic based on belief alone. It is also an extreme level of consistency programmed into the sorcerer's psyche which allows the sorcerer's activities to subtly influence the causality of the physical environment.

Every action we make has an effect on the physical environment. Only very specific actions can have a non-local effect. At this point, let us examine the rule that consistency sends energy to inconsistency. This is a statement of the propagation of simplicity based on Occam's razor. Occam's razor proclaims that the simplest explanation is the best explanation. We can extrapolate from this to deduce that given two systems, the simplest functional system is the best system. An inconsistent system will tend to be more complex than a consistent system. Thus, when two systems interact, due to the propagation of simplicity, the more complex system will assume a structure which emulates the simpler system.

These arguments are worth repeating in a different way.

The rule that consistency sends energy to inconsistency is an alternative description of the propagation of simplicity. It may seem to be at odds with the fact that disorder tends to increase, but the fact is there is a difference between disorder and complexity. Disordered systems are not complex. They are simple. It is very easy to create disorder. It is difficult to create complexity. The fact that complex systems are more prone to inconsistencies does not preclude the fact that a complex system may in fact be the more consistent system. However, given a complex system and a less complex system that each accomplish the same task, it is the simpler of the two systems that is the most efficient. If allowed to interact, the two systems will gravitate towards the most efficient functionality. Thus, simplicity propagates. It is clear that this rule has an affinity for the rule of evolution. That is, evolution depends on survival of the fittest.

Everything we do matters and every action we make is a reaction to previous actions superimposed with a measure of free will. Let us examine the concept of rational causality. Causality is the connective force underlying sequential actions which form a segment of the causal network underlying reality. When one action occurs, it occurs in a fashion which is a direct response to the resultant of causal forces created by prior actions. For inanimate matter and machines, the causal network of reality is deterministic at the macroscopic level. At the microscopic level, that determinism is diffused by the presence of uncertainty due to the properties of quantum mechanics. The perceptual experience of a human being is at the macroscopic level, so a human being perceives the activities of the physical universe as largely consisting of deterministic activity. However, a human being has the capacity to behave in a way which is not strictly causal because of her free will. Rational causality is the connective force underlying sequential actions which allow for a range of responses that fall within a standard deviation determined by the definition of rationality dictated by society. In rational causality the causal chain is rendered fuzzy by responsive actions which vary from the logically consistent responses required by strict causality. In simple terms, the freedom of the human being allows her to produce unpredictable actions which are bounded by the limits of rationality. When the human being exercises too much freedom, their actions are deemed irrational by society and their activity thus exists outside of rational causality.

Since we are conditioned by society, we find ourselves constrained by rational causality. It is only when some unpredictable phenomenon occurs that rational causality is broken. At that point, our freedom is unrestricted since the causal forces are momentarily nullified. It is in such moments of unlimited freedom that we find new meaning and are able to shape our destiny. That is, we are able to re-route the causal chains that determine the evolution of our life experience.

This is a very significant result. Unpredictable actions are the results of inconsistencies and inconsistencies are mistakes. We are saying that it is the mistakes in life which gives us the opportunities to mold our experiences. Nietzsche said that what doesn't kill us makes us stronger. It is interesting to note that the freedom of the sorcerer is limited by their level of consistency. At the times in the sorcerer's life when they are more prone to make mistakes and their decisions create more risk, they have the greatest potential for changing their path and affecting their physical environment. Of course, this applies to everyone. It is when we are the most vulnerable that we have the greatest impact on life.

The training required to do magic incorporates exercises in concentration, meditation, symbolic manipulation, and visual imagery. It also requires the practice of consistent conceptual products. That is, the sorcerer practices the construction of concepts which remain wholly consistent with respect to potential knowledge. The sorcerer avoids whimsical, inconsistent thoughts such that every thought sequence produced by the sorcerer has a logical substructure and/or coherency. The reason for this is the fact that every action has a consequence. This applies to every thought sequence. Fortunately, the standard deviation of possible responses to every situation which maintains consistency with rational causality allows for an intrinsic level of unpredictability. Consequently, there is an allowable variation to the level of logical consistency of a thought sequence. The evolution of the sorcerer's mind into a consistent state is aided by meta-duality, the superposition of extremes. In the science fiction series, Star Trek, the Next Generation, there is a character called Data who is an android with a positronic brain. We can borrow this terminology when discussing the Ω state created by meta-duality. The Ω state is a positronic state. The term positronic is apt, because the statistical interaction between positivity and negativity will ultimately be dominated by positivity. This occurs because positive themes are cooperative, while negative themes are disruptive. The positronic state exists beyond extremes and it is important that it is the result of repetitive interactions between positive and negative themes. That is, the positronic state must evolve with positivity as an emergent feature. This makes a very important point. When conditioning the mind, successful results will not occur if either positivity or negativity is specifically selected. The reason for this is the fact that duality requires the interaction of extremes such that isolation of an extreme diminishes meaning. In order to successfully move beyond duality, we must sustain the tension between extremes. However, if the superposition of extremes occurs as an external phenomenon, meta-duality can be dangerous. This would lead to activities which contain both pleasure and violence as the extremes manifest in the outer universe. Thus, for the sorcerer, meta-duality is an internal phenomenon and consequently, the experience of extremes remains purely mental.

At this point, we would like to review concentration. Concentration might be thought of as a mental focusing or thought pressure. When we concentrate, we examine a sequence of thoughts to determine their essential meaning and to reduce and reconstruct these sequences into related formats. While concentrating, we may move thought sequences into various contexts to examine their information content. The goal of concentration is to reduce the level of abstraction of a thought sequence. In other words, concentration blue-shifts or cools a thought sequence by reducing its frequency. Note that concentration is a property of consciousness as is meditation. Consequently, the energy dissipated by concentration heats consciousness while the thought sequence which is the focus of concentration cools down. The fact that meditation cools down consciousness is the reason why meditation is often combined with concentration to allow it to maximize without overheating consciousness. Concentration largely involves logical thought. Thus, it is largely a function of the left brain.

Symbolic manipulation generally consists of some language or some form of equations which emulate the internal relationships of a thought sequence. To a large extent, symbolic descriptions or equations represent a synthesis or model of the given thought sequence. In that sense, symbolism or equations correlate to a reconstruction of the essential meaning of the thought sequence. It is the synthetic element of formulating symbolic descriptions which makes the process largely inductive. Induction is a property of the right brain. We observe that this situation depends on the originality of the symbols. They must be the result of a purely creative process in order to be the product of the right brain. Alternatively, when we use pre-existing language to solve problems, we are performing a reductive process and are thus utilizing the left brain.

Visual imagery is normally a property of diagrams and drawings, but it might translate into audible imagery as well. Thus, we might satisfy the requirements of visual imagery by using music instead of art. In addition, the imagery might be in the form of language.

Again reviewing the essential theory of magic, we start with concentration to blue-shift a thought sequence which we want to manifest. We then produce a spell or a code which encapsulates the original thought sequence in the form of symbolism and/or imagery. We want the code to be understood to such a level that it is nearly instinctual. The code by its very nature and the intrinsic energy created by the process of concentration then radiates to a receiving matrix, which is the physical system in which the original thought sequence manifests. The only case in which there is no spell is that in which concentration is in the form of belief. Belief is only effective as a transmitter of magic when the inner universe is more consistent than the outer universe.

We have identified a master of magic as a sorcerer, consistent with traditional terminology.

An essential principle of magic is the fact that concentration transforms a thought from an imaginary state to a real state. Since the magnitude of imaginary energy is significantly greater than the magnitude of real energy, there is a residual energy of transformation that becomes available for transmission of the thought sequence through real space. Consequently, the effectiveness of a spell depends on how much the original thought sequence is shifted in frequency.

Another essential principle of magic is the propagation of simplicity. When magic is based entirely on belief, it is transmitted by the tendency of simplicity to propagate to more complex systems. A consistent inner universe implies the simplest possible configuration, given the needed functionality. Thus, the system is minimally complex. Because themes are normally compartmentalized in the inner universe, the human psyche is typically a very complex system. This is because the themes may sustain inconsistent logical substructures while separated. The evolution of the inner universe promotes consistent logical substructures throughout the system, which naturally minimizes the necessary complexity.

Magic occurs when we are able to completely understand a symbol, emotion, intuition, or fantasy. When that understanding becomes instinctive, magic is created. From this perspective, it is clear that magic occurs everyday. You don't have to be a sorcerer to perform magic. The difference is that the sorcerer's magic is intentional, while our magic may be happenstance.

It is important to consider how one defends oneself from magic. Based on the description of magic given thus far, it is reasonable to believe that a complex network of magical disturbances happens all the time. The magic, of course, is typically disguised as traditional forms of energy. A spell transmits energy to a matrix and a human being may often be part of that matrix. The effectiveness of the spell depends on how well the matrix causally connects to its physical environment. For purposes of simplicity, consider magical defense as isolation of a matrix. In essence, we want the matrix to be causally isolated from physical reality. In that eventuality, there is no way for a spell to causally connect to the matrix. That is, energy impingent on the matrix cannot logically connect to the substructure of the matrix. This isolation may be achieved by encoding the matrix such that its internal information is not recognizable by a traditional knowledge network. Successful encoding implies that there exists a consistent internal mapping between the internal and external logical substructures, but that mapping resides within the matrix and is not easily available. Only the intelligence within the matrix understands the mapping or copies of the mapping may exist in hidden locations. This encoding is approximately a private language. A perfectly private language is impossible, but a heavily isolated language is not. If a language were perfectly private, there would be no mapping and thus no way for the intelligence to communicate with the outer universe. When an intelligence cannot communicate, it effectively cannot exist. An encoded matrix is basically a reflector of energy.

A physical system can either reflect, absorb, or transmit incident energy. When a matrix absorbs the energy of a spell, the original thought sequence is manifested. Consequently, an alternative way to avoid magic is by transmitting the spell's energy through the matrix. Transmission occurs for a rigid matrix. That is, the logical substructure of the matrix is precise such that there is no standard deviation of behavior. The rational causality of the matrix becomes purely causal, because there is no rational flexibility. In that case, the matrix will simply transmit the incident energy of the spell into the physical environment. An example of a rigid matrix would be a rock. Thus, it is very difficult to harm a rock with magic. Manifestation can affect any system that is not rigid or has degrees of freedom. A degree of freedom indicates a probability of potential behavior. Simple machines have few degrees of freedom, whereas a computer has many degrees of freedom. Consequently, manifestations with computers occur more readily than manifestations with simple machines. Exceptions to this occur when there are inconsistencies in the machines. In that case, a manifestation can cause the machine to malfunction.

It has been theoretically demonstrated in the book, *Metaphorms and Information Physics* that intelligence can be linked to probability distortions. Since intelligence is a requirement of concentration, it follows that the probability distortions created by intelligence represent magical phenomena. To the extent that intelligence is competitive, we find that the manipulation of one intelligence by another can often be due to magic.

Most magical phenomena are consistent with the known laws of physics. It is only on rare occasions that magical phenomena extends itself into unknown regimes of behavior. From the perspective of quantum mechanics, a physical law is only a highly probable event. A highly improbable event can counter a physical law. Magic generally deals with the improbable which maintains consistency with the laws of physics. However, on rare occasions it is possible that a highly improbable event manifests which counters those laws. Such magic would be called miraculous.

Essentially, magic is the physical activity which through causal connectivity creates remote disturbances which have specific consequences. It is clear that physical reality is a network with causal connectivity. The challenge is to properly disturb one region of reality such that a remote region experiences specific activity. The solution is to condition the mind to send specific signals to the body which in turn perturbs the local physical environment in a very precise manner, transmitting specific energy at specific times. This causes a chain of signals which propagate to remote locations and activates physical systems in a specific manner.

What you think about determines how you act. Your actions have reactions which can trigger a series of activity until that activity propagates to some remote destination. However, since reality is such a vast, complex network, it is extremely difficult to determine which actions at this point in space affect other actions at some other point in space.

The purpose of the theory of magic is to condition the mind in precisely the manner needed to affect local activity which will garner an appropriate response from some remote location. We have proposed that it is the thorough understanding of a thought sequence which produces the symbolic and/or graphic structure that causally transmits energy to a specified matrix. We call that symbolic and/or graphic structure a spell. The amount of concentration needed to produce and/or understand the spell is a measure of the energy transmitted by the spell.

We can identify the driving force behind the effectiveness of magic as the principle of complimentarity of concept and representation. That is, as the number of concepts grows, the number of associated representations reduces. Alternatively, as the number of concepts reduces, the number of associated representations grows.

The spell is a representation of a concept implicit in the original thought sequence. Initially, in the first stages of concentration, there are multiple concepts associated with the thought sequence as the ideas are not consolidated. Consequently, there are few associated representations. For example, these initial representations may consist of models, notes, and diagrams. However, when the thought sequence is fully consolidated, the concept becomes unified and the principle of complimentarity compels physical reality to contain an infinite amount of associated representations. The ensemble of these representations coincide with the matrices designated by the spell.

A single valid and robust concept will completely radiate as representations within physical reality. However, the representations will be the most prominent in the closer matrices. By selectively reducing the distance between a spell and a matrix, a sorcerer can ensure that energy is transmitted to the matrix. The energy radiated by a spell is spherical with an intensity proportional to one over the distance squared. Therefore, a spell generates a scalar field with no particular direction. We reiterate that spells with specific objectives are cast such that the distance between the spell and the matrix is minimized.

When a sorcerer is purely in a state of meditation, when the body has a completely predictable motion and the mind produces no intentional thought sequences, there is no energy radiated by magic. The sorcerer then generates a null field. Otherwise, when not in a complete state of meditation, the sorcerer will continually radiate magic based on the presence of consolidated thought sequences in his mind. The more a sorcerer thoroughly understands, the more magic he will radiate. However, intrinsic spells within the mind of the sorcerer are multiplexed by his consciousness and thus, the intensity of their radiation depends on the focus of the relevant temporal subintervals. That is, magic will radiate the most from the ideas which are the most concrete and which hold most of the sorcerer's attention.

We can ascribe to a spell the attribute of density which is a measure of the localization of the spell. That is, a very localized spell is only privately known. Thus, while a high density spell is only privately known, a low density spell is well known. The more dense a spell is, the more it radiates. That is the reason why well known ideas generate minimal magic. The sorcerer may retain many dense spells and thus continually radiate magic. Note that the presence of radiation from magic implies that energy disturbances due to probability distortions ensue. Consequently, the existence of a sorcerer warps the probability distribution of macroscopic reality. In order to compensate for this effect, the sorcerer can persistently meditate, thus reducing the radiation field. Alternatively, the sorcerer may intentionally reduce the density of his spells by sharing them with others.

It follows that encoding the sorcerer's matrix increases radiation, because with encoding, the density of spells increases. Thus, when the sorcerer is highly protected from external magical influence, he is also an enhanced source of magic.

While probability distortions created by magic will reroute causal activity within a matrix, they will also cause the matrix to malfunction in partitions that are inconsistent. Since magic radiates uniformly in all directions, all matrices in the vicinity of a sorcerer are subject to distortions.

We find this to be relevant to the nature of intelligence. As intelligence grows, it becomes unambiguous, but as it continues to grow, its appearance shifts and it becomes increasingly undetectable. However, intelligence continues to distort probability.

There is the question of good and evil with respect to magic. The nature of magic is a reflection of the nature of consistency of the associated sorcerer. If the consistency of the sorcerer is such that activities which can be morally described as good predominate, then the magic produced by the sorcerer would be characterized as good. On the other hand, if the consistency of the sorcerer is such that activities which can be morally described as evil predominate, then the magic produced by the sorcerer would be characterized as evil. Since the stable existence of a sorcerer is correlated to the nature of the outer universe, evil magic would require that some consistent interpretation of the outer universe exists in which evil predominates. From the perspective of duality, we cannot say that reality is either good or evil, since its nature depends on our perspective. However, from the perspective of meta-duality, the proposition that with the superposition of extremes, positivity reinforces itself while negativity diminishes itself, given repeated interactions supports the perspective that there is ultimately a positive overtone to reality.

The distribution of perspectives about reality is a direct consequence of the processing architecture of the associated intelligence. We have previously divided the possible architectures into three primary configurations, dual, non-dual, and meta-dual. In a dual architecture, there is a selective preference of one extreme over another extreme. In a non-dual architecture, there is no distinction between extremes. In the case of a meta-dual architecture, the extremes are simultaneously equivalent and distinct. The expression of intelligence for these different architectures varies and this has consequences for the development of a sorcerer. A dual thinker exhibits reductionist thinking. A non-dual thinker exhibits holistic thinking. A meta-dual thinker exhibits both reductionist and holistic thinking. Reductionist thinking consists of analysis and decomposition and consequentially is more coherent with negativity or the removal of substance. Holistic thinking consists of synthesis and composition and consequentially is more coherent with positivity or the addition of substance. It follows that a meta-dual thinker incorporates both the positivity of inclusion along with the negativity of exclusion as a superposition.

In the interest of simplicity, we shall refer to both an intelligence and an arbitrary physical system as a matrix. Thus, magic can be thought of as interactions of matrices via probability modifications propagated by energy. The consistency and encoding of a matrix are a measure of how well it radiates energy and how shielded it is from external radiation. From this simpler perspective, there is no need to discuss spells since they are an intrinsic property of a matrix. Furthermore, we submit that the level of encoding of a matrix correlates to the density of the matrix. Concentration compresses a matrix and the rigidity of the matrix is a measure of the energy dissipation due to concentration. Cool matrices are highly compressed, while hot matrices are highly expanded. The temperature of the matrix refers to the frequency content of its logical substructure.

We note that the consistency of a matrix is a measure of its causal integrity. In general, the more rigid a matrix, the more consistent it is. Corrections of inconsistencies enable the disruption and redirection of causal chains. This follows from the ideas presented on rational causality. Magical influence essentially modifies causal chains in a matrix. Thus, consistency naturally inhibits such influence. We can extrapolate that in a completely consistent reality, magic would be ineffective.

The interaction of matrices which characterizes magic involves the transmission, absorption, and reflection of probability waves. It is only through energy exchanges between the matrix and the environment that most probable activities are stimulated. Thus, a probability wave cannot affect a matrix which is energetically inactive. The fact that magic incorporates probability waves disconnects the magic from the causal chains which it influences. Thus, there is never any definitive evidence that magic exists.

The Theory of Probability Distortions

At this point, we can redefine magic as a probability distortion. Probability distortions occur when matrices interact by exchanging probability waves. A sorcerer, equivalent to a matrix, radiates probability waves. Consider that a matrix is a three-dimensional configuration of time-dependent probabilities, designated with the subscripts, i, j, and k which correlate to a specific three-dimensional logical substructure, designated by the superscript, l. The variable is thus a time-dependent probability for a subcomponent of a specified state and is given by

$$p_{ijk}^{(l)}$$

We can designate a probability distortion as a perturbation given by

$$\delta p \,.$$

Consequently, assuming the presence of energetic disturbances, the original matrix undergoes a transformation from the lth state to the mth state given as

$$p_{ijk}^{(l)}(t) + \delta p \rightarrow p_{ijk}^{(m)}(t + \delta t) \,. \tag{1}$$

This transition designates the most likely state after a probability distortion occurs, but we emphasize that if the matrix being perturbed is not exposed to energetic disturbances, then the new state will not occur. Essentially, the matrix must undergo motion in order for the more probable configuration to emerge. It is clear that the interactions of matrices are not causally linked.

The probability distortion is an impulse which can be converted into energy by utilizing the equation for the energy contained in an information rate, I. That is,

$$I = -\hbar\omega\log(p)_,$$

(2)

where $\hbar$ is Planck's constant divided by two pi, ω is the frequency of the symbols in the information sequence, and p is the probability of the information. The equation (2) basically says that the rate at which information changes is proportional to the rate of change of a symbol times the self-information in a symbol.

The equivalence of information rate to energy can be explained as follows. Consider a clock with 99 percent of its mass along its perimeter. The time evolution of the clock occurs via a small motor of negligible mass which rides the clock perimeter. A position of the motor designates a time value. Consequently, two time values correspond to a fractional mass of the clock. If we encode the time values into symbolic information such that the distance between symbols is proportional to the difference in information presented by the symbols, then a change in information is equivalent to a fractional mass of the clock. Since mass is equivalent to energy, we can conclude that a change in information is equivalent to energy. Thus, information rate is equivalent to energy.

It follows that the energy impulse created by a probability distortion is thus

$$\delta I = -\hbar\omega\log(1-\delta p)_,$$

(3)

in which it is clear that the probability distortion is non-negative. That is,

$$0 \le \delta p < 1$$
.

We have previously stipulated that the radiation of a matrix is inversely proportional to the square of the distance between interacting matrices. Therefore, the intensity of a probability distortion reduces with distance. Designate the probability distortion at its source as

$$\delta p_0$$
.

If r is the distance traveled by the distortion, and r_0 is the position of the source matrix, then we can write

$$\delta p = r_0^2 \frac{\delta p_0}{r^2}$$
.
$$(4)$$

Using the three spatial variables, x_1, x_2, and x_3, we can write the spatial distance as

$$r = \sqrt{x_1^2 + x_2^2 + x_3^2}$$
.
$$(5)$$

The energy impulse carried by a probability distortion is extremely small as evidenced by equation (3) because Planck's constant is very small. We propose that a significant energy impulse is the result of repeated distortions. That is, probability waves create sequences of probability distortions at a rate consistent with their frequency. Each distortion is an independent impulse which accumulates at a receiving probability distribution. We can designate an individual probability distortion with an index n.

$$\delta p \to \delta p_n, n \in \{1, 2, ...\}$$
.
$$(6)$$

The total impulsive energy is the change in information rate:

$$\Delta I = -\hbar w \sum_n \log\left(1 - \delta p_n\right)$$

$$(7)$$

It follows that the sequential probability distortions depend on the distance they individually travel, r_n:

$$\delta p_n = r_0^2 \frac{\delta p_0}{r_n^2}$$

$$(8)$$

Consequently, the impact of probability waves is statistical, requiring a coherent sequence of probability distortions. This is consistent with the notion that successful manifestations require focus and time. We assume that there is a mechanism within the receiving matrix for converting the energy deposited by the probability distortions into the original distortions:

$$\Delta I \rightarrow \sum_n \delta p_n$$

This result alters equation (1) to

$$p_{ijk}^{(l)}(t) + \sum_n \delta p_n \rightarrow p_{ijk}^{(m)}(t + \Delta t)$$

$$(9)$$

in which the total time interval is the sum of the interval of each distortion:

$$\Delta t = \sum_n \delta t_n$$

$$(10)$$

Equation (9) designates that a state change in the matrix is likely after multiple probability distortions. At this point, we can specify that in the far-field, a probability wave is a spherical wave given as

$$u_n = r_0 \frac{\sqrt{\delta p_0}}{r_n} e^{i\left(\omega t - \vec{k} \cdot \vec{r}_n\right)} \quad , \tag{11}$$

where k is the wavenumber:

$$k = \frac{\omega}{c} \quad , \tag{12}$$

with c being the speed of light, which is the speed of the probability distortions.

Based on interference, the net probability wave can be written as

$$u = \sum_n u_n \quad . \tag{13}$$

Using the designation

$$\delta p = \sum_n \delta p_n \quad ,$$

for the cumulative probability distortion, we find that δp is the total "energy" of the wave:

$$\delta p = \sum_n u_n^* u_n \quad . \tag{14}$$

The "energy" carried by a probability wave is in fact probability. The true energy of the wave is given in equation (7).

The mind multiplexes through all possible thought sequences. Each sequence has a certain probability which determines its information content. The total probability is unity. We can designate this as

$$\sum_i p_i = 1 \quad . \tag{15}$$

If we divide the thought sequences into foreground sequences and background sequences, then concentration is focused on the foreground sequence which has probability p_0. Conservation of probability for foreground and background is thus

$$p_0 + \sum_{i=1} p_i = 1$$

$$(16)$$

When concentration occurs, there is a natural oscillation between foreground and background sequences. This creates a distortion given as

$$\delta p_0 = \sum_{i=1} p_i - p_0$$

$$(17)$$

The less probable the foreground thought sequence, the greater the distortion and the more consolidated the foreground sequence, the more uniform the distortion.

As the mind multiplexes, the distortion will oscillate, causing radiation. Clearly, if the foreground thought sequences are undifferentiated from the background sequences, as in meditation, there is no distortion. It also follows that only weak distortions occur if foreground sequences are highly probable.

The fact that a less probable foreground thought sequence is relatively complex is consistent with the previous proposition that concentration on higher frequency thought sequences produces greater transmission of the associated manifestation. This follows since the greater the probability distortion, the greater the radiation.

Prior emphasis on the importance of visualization is consistent with the need to tie probability distortions to a targeted probability distribution as indicated in equation (9). We refer to this visualization as tuning. As an example of the importance of tuning consider the following. A person is pacing within a room. Consequently, l designates a position state. The room is divided into three regions, the left, the middle, and the right. Over an extended period of time, the person will spend different fractions of their time in the three regions as they pace the room. This is likely due to the time they spend idling or the different rates of motion of their pacing. We can specify the original probability distribution, $p^{(l)}$ as

$$p^{(1)} = (0.1 \quad 0.3 \quad 0.6).$$ (18)

Clearly, the person spends 10% of their time on the left side of the room, 30% of their time in the middle of the room, and 60% of their time on the right side of the room. Suppose that we would like them to spend most of their time in the middle of the room with probability distribution, $p^{(m)}$, so we send a cumulative probability distortion, 0.3, tuned to the middle of the room.

$$p_i^{(l)} + 0.3 \rightarrow p_i^{(m)}.$$ (19)

The tuning implies that i=2.

There are two possible distributions for $p^{(m)}$:

$$p^{(m)} = (0.1 \quad 0.6 \quad 0.3)$$ (20)

or

$$p^{(m)} = (0.0 \quad 0.6 \quad 0.4).$$ (21)

Thus, after the cumulative probability distortion, the person spends 60% of their time in the middle of the room.

Suppose the distortion was improperly tuned so that it was directed towards the left side of the room. In that case, we would have

$$p^{(l)} \to \left(0.4 \quad s \quad r\right),$$

(22)

with $s + r = 0.6$.

As a result, there is no guarantee that $p^{(m)}$ is achieved.

It is important to note that the frequency, ω of a thought sequence is not a measure of how we experience the thoughts. Rather, it is a measure of how fast the mind multiplexes.

We want to point out that the information content of a thought sequence can be maintained even if the thought sequence becomes instinctual. That is, even if a thought sequence is improbable, we may still retain an instinctual understanding of it. For example, one may not ride a bicycle for many years, but when the opportunity arrives, the ability to ride the bicycle is easily retained. One may have a thorough understanding of physics, but it remains a complex subject. This supports the uniformity of a probability distortion. Understanding doesn't alter information content. Rather it is the frequency of occurrence which alters information content.

Referring to equation (9), observe that when the matrix either reflects or transmits the probability distortion (or its associated energy signature), m=l and there is effectively no state change.

A state of a matrix can be many things. It may be a person riding a bike or a computer running a program or a car moving along a road. It could also be a person solving a problem. The situation is completely diverse. The difference between one state and another state is determined by the difference between the causal paths of the states. If either an error or an unpredictable event occurs, then the causal path will change in response to the occurrence. This occurrence is clearly a probability distortion. The modification of causal paths is metaphoric to Newton's first law of physics. That is, a body at rest or in motion will remain so until acted on by a force.

It follows that the larger the motion of the body, the larger the force needed to change that motion. Similarly, the larger the energy of the causal sequence, the larger the error or the more unpredictable the event needed to alter the causal path.

A causal sequence in the mind is a thought sequence, which is a current. There is a resistance to altering a current which is a statement of Lenz's law in physics. We can designate the quantity of current as a measure of the complexity of a thought sequence. Consequently, complex thought sequences resist change while simple thought sequences defer to change.

Because the thought sequences of a sorcerer are heavily encoded, they will tend to be complex and therefore resistant to change. This suggests that the sorcerer is less subject to probability distortions. Since inconsistency creates complexity and the sorcerer's mind approaches consistency, we emphasize that the complexity of the sorcerer is always relatively minimal. The simplest, effective configuration may be the least complex, but it may still maintain complexity. Due to consistency, the sorcerer's mind will tend to be less complex than the environment. This will create a force due to the propagation of simplicity, that will radiate probability distortions from the sorcerer's matrix.

In this context, we must differentiate between inconsistency and randomness. While inconsistency can still accomplish work, randomness cannot accomplish work. Thus, randomness is simple. For example, consider an extremely inconsistent theory like the earth-centered solar system. It contains many complex geometric patterns to explain the orbits of the planets and the sun around the earth. It requires deep concentration and mental effort to understand this theory. Consequently, there is much work involved. The correct theory is the heliocentric theory in which the planets orbit the sun. It is much more elegant and less difficult to understand. This is evidence of the power of simplicity. Now consider the description of the solar system given by an imaginative five-year old. There will generally be no logical structure to such a description and although it will be seemingly complex, it is actually random and there is really no work needed to appreciate it. Thus, we can conclude that such a description is simple.

Recall that we stated that a probability wave cannot affect a matrix which is energetically inactive. It follows, that a probability wave cannot give life to inanimate matter.

As a sorcerer evolves, he becomes less subject to entertaining unfounded thought sequences. Thus, entertainment for a sorcerer consists of realistic scenarios or consistent speculation.

We have dismissed with the explicit notion of a spell and accord it implicitly to the structure of a matrix. Recall that a spell is essentially a code. We might think of a code in this context, as the language of the matrix. What we once referred to as a dense spell thus refers to an encoded matrix. For intelligent matrices, encoding is equivalent to compression of a thought sequence and thus, concentration is proportional to compression. It has been proposed that information is proportional to length. Therefore, concentration reduces the length of a thought sequence, resulting in a compressed or encoded sequence of reduced length. The information gained through encoding compensates for the information loss in length.

We might think of concentration as the methodology for creating irradiative sources of probability waves. Concentration is maximally effective when combined with meditation as a cooling process. This supports meta-duality as a primary tool of the sorcerer.

It has been shown that probability distortions produced by matrices induce remote physical phenomena or manifestations. The critical involvement of probability implies that a manifestation cannot be causally linked to a source. This also creates a constraint for the distribution of manifestations. The reason is because the sum total of probability is unity. Consequently, when probability distortions augment probabilities in one regime, they necessarily reduce probabilities in another regime. Thus, when some individuals succeed at manifestation, other individuals must fail.

It behooves the sorcerer to maintain a high level of creativity. That is, the sorcerer should incorporate unique thought sequences which correlate to unique physical phenomena. This augments the diversification of reality and avoids competitive manifestations.

Alternatively, when a sorcerer focuses on the mundane, probability waves create distortions which can disrupt normal physical phenomena. Mundane phenomena are the most commonly thought of phenomena. We consider concepts like power, fame, and fortune to be the mundane.

A sorcerer can create a source of probability waves as a set of symbolic patterns. This may be anything from a book to a diagram to music. In the earlier discussions of magic, this was referred to as a code or a spell. Probability distortions will be created by this source in the same way that they originate from the sorcerer. This is similar to the manner in which objects exposed to radiation will reradiate on their own. Consequently, we will hence refrain from using the term sorcerer and refer only to a source which means a source of probability waves.

A source produces probability distortions that induce a probability distribution in a receiver which is most likely to generate a desired physical state. The transition to the desired state occurs after stimulation by local energy disturbances. This transition has been called a manifestation. Manifestation is not possible if the original probability distribution of the receiver is unknown. The reason for this is because probability distortions must be tuned to the receiver. This requirement of tuning explains why manifestations for misunderstood systems are rare.

It has been pointed out that meditation generates a null radiation field of probability waves. This follows because in meditation, foreground thought equals background thought. Thus, consistent with equation (17), meditation produces no probability distortions.

We point out that in our theory we do not use probability densities because we are referring to discrete systems. The theory could readily be written in terms of probability densities if continuous systems are considered.

Quantum mechanics is a theory which predicts probability densities for physical states. It recognizes that changes in probability densities are a result of energetic disturbances, but it doesn't recognize that the energetic disturbances can be caused by probability distortions.

Since probability distortions can be caused by the mind, this theory proposes a mechanism for converting thought into energy. Consequently, it is a solution to the mind-body duality. That is, the mind controls the body via probability distortions. In fact, it suggests that the distance, r, between the mind and the brain is infinitesimal, leading to minimal divergence of the probability waves, which, in turn causes a maximal energy transmission due to probability distortions. This maximal energy transmission is apparently above the critical level for mind to body manifestations. The total impulsive energy of the mind follows from equation (7) as

$$\Delta I = -\hbar\omega\sum_{n}\log\left(1-\delta p_{0}\right) \tag{23}$$

Manifestation is straightforward. In order to manifest a thought sequence, we must concentrate on that sequence until it is thoroughly understood. That understanding produces a probability distortion that transmits through a causal chain of systems until it is absorbed by the targeted system.

Because manifestation leaves no causal physical evidence, it might be considered to be subjective or coincidental. However, there is actually no such thing as coincidence. Everything happens for a reason. In fact, the test of manifestation is the measurement of concentration and its effectiveness.

As a point of clarification, there is an energy associated with communication, given by equation (2) and there is an energy associated with concentration given by equation (3). We are particularly concerned with the energy associated with concentration since it leads to manifestation.

The infrequency of manifestation follows from the fact that many systems have rigid probability distributions not subject to distortion or their distributions are too complex to be understood. When a system cannot be understood, a source cannot tune in to it.

Manifestation includes both the left brain and the right brain if we consider, decoding and encoding of the original thought sequence, respectively. However, the energy dissipated in concentration is primarily due to decoding since encoding is essentially a creative process. It is similar to the fact that science requires more mental effort than art. The left brain determines the complexity of a source, while the right brain determines its originality.

If not properly tuned, probability distortions can maximize inconsistencies, causing failure or error. It is reasonable to surmise that the reason for so much disorder in the world is due to improper tuning of probability distortions.

Probability distortions produce very little energy. It is only enough energy to stimulate energy exchanges in the environment. Consequently, the conversion of thought into energy is not a viable source of energy. Essentially, manifestation is the redistribution of energy, not the production of energy.

Manifestation is not simple wish fulfillment. If we want to be rich, we have to understand exactly how it is that we can be rich.

If we want to be famous, we have to understand exactly how it is that we can be famous. If we want to be smart, we have to understand exactly how it is that we can be smart. Probability distortions will connect potentiality to realization.

If our conclusions about a thought sequence are based on belief rather than concentration, then our belief can also produce probability distortions. However, the intensity of those distortions depend on the strength of our belief. When our belief system is extremely consistent relative to the outer universe, manifestations will occur.

Given a radiating source, local activities stimulated by probability distortions will trigger energy disturbances which propagate coherently towards sites which are remotely affected by the distortions.

With respect to the characterization of magic with the use of different spells, in particular, an X-spell, a Y-spell, a Z-spell and a Q-spell, we find that these correlate, respectively, to greater and greater probability distortions.

The law of attraction suggests that like attracts like. The theory of manifestation is consistent with this law. Concentration in a source configures the source to be consistent with a particular thought sequence. Probability distortions generated by the source stimulate remote sites to be similarly configured.

There is a big difference between manifesting a one-time event and manifesting a permanent phenomenon. In the first case we alter probability. In the second case we alter reality. In the latter case, manifestation becomes science. It is probable that some of the greatest scientific discoveries were manifestations.

When we manifest a thought sequence, that sequence develops a precise meaning. It is because of that precise meaning that representations of that thought sequence imprint themselves onto reality. The total probability of background thought is statistically stable. Because of its precision, the probability of the foreground thought is also stable. These probabilities form a perfect dipole about their center. The oscillation of multiplexing causes these probabilities to radiate probability waves. The radiation does not occur for normally fuzzy foreground thought.

The type of meditation discussed in this book is the form which ignores thought. As we have shown, that type of meditation does not manifest because foreground thought equals background thought. However, there is a form of meditation which focuses intensely on single objects. In that case, there is also no manifestation because the object of meditation already exists.

In manifestation, we focus on things which don't exist and we bring them into existence. Essentially, we transfer things from imagination into reality. Given an object of manifestation, we form it as completely as possible within our imagination. At some critical level of structure, the object transfers into reality. In other words, a perfectly realistic story will transform from being fiction into being non-fiction.

At this point in the development, the physics of manifestation is effectively summarized by the equation:

$$p_{ijk}^{(l)}(t) + \sum_n \delta p_n \rightarrow p_{ijk}^{(m)}(t + \Delta t)$$

,

in which $p_{ijk}^{(l)}(t)$ is the original probability distribution of a remote site, $\sum_n \delta p_n$ are the cumulative probability distortions, and $p_{ijk}^{(m)}(t + \Delta t)$ is the final probability distribution of the site. The designation of specific physical states is indicated by the superscripts which give the lth state as the original physical system and the mth state as the final physical system. Observe that due to conservation of probability, we must have

63

$$\sum_{i,j,k} p_{ijk}^{(l)} = \sum_{i,j,k} p_{ijk}^{(m)} = 1$$

$$(24)$$

When highly specific tuning is considered, an even simpler equation for manifestation is given as

$$p_1 + \delta p = p_2,$$

$$(25)$$

in which p_1 is the initial probability and p_2 is the final probability. In this case, we identify δp as a cumulative probability distortion. In general, the distortions may directly affect multiple probability components of the distribution $p_{ijk}^{(l)}$, so equation (25) is a special case.

Let's review the steps of a manifestation. We begin with concentration on a thought sequence. Concentration is maintained until the thoughts become focused in the foreground of the mind. Focus may be enhanced by physical representations of the thoughts. The attentiveness to those foreground thoughts is naturally multiplexed with background thoughts. This causes an oscillation that produces probability waves with an "energy" equal to the probability distortion between background and foreground thoughts divided by the distance squared. Thus, the probability waves are essentially spherical and diverge as they radiate. The probability distortions carry real energy which is absorbed at a specific site designated by its attributes as a receiver. Based on tuning, the energy carried by the probability distortions is decoded into an appropriate distortion of the site's probability distribution. The subsequent behavior of the site will be consistent with the altered probability distribution and is considered to be the desired manifestation. We add that the manifestation may be augmented by the superposition of energy disturbances propagated to the site from the source of the probability waves by local source activity and the causal connectivity between the source and the site.

Consider the following example of a manifestation. A source radiates probability waves which are tuned to send distortions to a nearby phone. The distortions increase the probability that the phone will ring. When the phone receives the probability distortions, their energy stimulates the circuitry in the phone. This is equivalent to the decoding of the energy into the original probability distortion. Stimulated by small disturbances like local air currents and ambient sound and light, the phone rings.

Let's consider another example of a manifestation. A source stands near a man and sends probability waves which are tuned to send distortions which increase the probability that the man will bend over. The distortions are absorbed by the man and their energy triggers a part of the brain which creates an urge for the man to bend over. This corresponds to the decoding of the energy. Stimulated by local energy disturbances, the man bends over.

We note that in these simple examples of manifestation, we have not included the more realistic requirements of time, trial and error normally needed to produce a successful manifestation.

The superposition of communicated information and probability distortions explains why some people are inordinately persuasive. Their communication is being enhanced by a manifestation of probability distortions in the mind of the receiver.

A manifestation can only perturb a site into an activity that is possible for it to do. That is, a manifestation only distorts existing probabilities. In the previous example of the phone being made to ring, it is clear that a phone commonly rings and thus, the manifestation is possible. Alternatively, if a manifestation was attempted in order to make the phone fly, then the manifestation would be essentially impossible because phones don't fly.

The existence of probability waves explains the connectivity between the mind and the body. Essentially, the brain interprets probability distortions from the mind. In turn, the brain controls the body.

Manifestation has a tendency to cause inconsistent systems to fail. This relates to the belief that certain theorists cause machinery to malfunction.

Consistency in a source will enhance the radiation of probability waves because of the propagation of simplicity.

Concentration

We would like to describe a form of concentration which causes the most salient themes to manifest.

First think, then observe your thoughts. Finally, perform some activity based on your observations. When you initially think, your thoughts will tend towards some interest, desire, intuition, or fantasy. Coalesce your thoughts around that attraction until you form a coherent thought sequence. The next thing to do is study that thought sequence. That is, define the terms, relate them to other sequences, and determine the consequences of those sequences. You will then want to do something about what you've learned, such as write, draw diagrams or some other associated activity. The representations you create will stimulate more thinking. At this point, you are ready to repeat the process. The thinking will be more focused, as will the observations, as will the activity. This is a method of concentration which utilizes the TOA information engine. That is, Think (T), Observe (O), then Act (A).

When you think with no a priori themes, your mind will consider local perceptions, memories, incidental problems and issues, and lingering desires. The mind will gradually focus on one of these items until it is clearly represented as a thought sequence. Perceptions of immediate physical phenomena will be considered. The more energetic phenomena will draw the most attention. As the mind adjusts to the energy level, it will be able to notice more subtle phenomena. Memories will be triggered by these perceptions. Some of these memories will be of recent events and some will be of more distant events. Due to these memories, outstanding problems will come to mind. Related issues will emerge. Tensions associated with lingering desires will be felt. All of these things form a cluster until some causal thread tightens and brings together an emergent theme.

The theme as a coherent thought sequence is ready to be studied. The study includes the definition of terms. That is, the examination of individual thoughts is carefully performed. This sequence is then carefully compared to other thought sequences. The meaning and impact of these thoughts on other thought sequences is considered. Possible replacements and extensions are evaluated. Ideas build and a pressure develops until the need for physical expression manifests.

Activities which physically represent the formulated thought sequence include writing, speaking, drawing diagrams, and compiling equations. These representations are meaningful constructs which naturally stimulate more thinking.

This represents the second cycle of the concentration process. At this stage there exists an a priori theme. Consequently, thinking is more focused and orbits about the given theme. All thoughts will have a common center and will be closely related. The mind will drift towards the equilibrium point of the various thoughts. The equilibrium point represents the balance of the thoughts which creates the new thought sequence to be studied.

The study of this new thought sequence is of higher resolution than the first study. The diversity of meaning of each thought is minimal since the meaning of the sequence is highly localized. The alternative thought sequences considered are sparse. This affords a higher frequency of consideration with fewer topics. A form of resonance promotes a need for representation.

Energetic imagery manifests this new thought sequence. The physical representation is finely attuned to the thought sequence. Written words will be relatively succinct. Equations will be precise. Diagrams will be specific. Articulation will be eloquent. This manifestation will produce more thought.

Thus, we continue with the third cycle of concentration. The orbital of thought has tightened and meaning is further restricted. Clearly, as this process continues, meaning becomes highly focused and a consolidated thought sequence emerges.

Causality

We live in a causality in which current events are caused by prior events. The junction between prior events and current events is called a causal nexus. We find that a rational individual has the ability to deviate from causality as long as the deviation has a rational explanation. This is referred to as rational causality. When the deviation from causality becomes irrational and there is no explanation for the behavior of the individual, then causality does not diverge from that causal nexus. In effect, that causal nexus no longer does work in the progression of reality. This is consistent with the fact that an entropic system is incapable of doing useful work.

It is a requirement that a manifestation be part of a rational causality. However, we find that the more improbable the manifestation, the more encoded the argument for rationality becomes. In other words, in order for a manifestation to be valid, it has to be supported by logical evidence. Nonetheless, the complexity of the evidence is a direct measure of the improbability of the manifestation.

In macroscopic reality, we refer to the deviation from causality as free will. Consequently, a manifestation necessarily requires free will because there are no probability distortions in causality.

Because it is incapable of producing consolidated, well-supported thought sequences, irrational behavior does not produce successful manifestations. This does not preclude the possibility that a manifestation may have begun with a thought sequence which was irrational in its original form.

As causality diverges, the thread of causality diffuses so that activities become increasingly independent. This is an area of caution for an individual, because rationality can be subtly lost. It is important to maintain a rational grounding.

In order to maintain the integrity of causality, it is important to remember that every action we take in this moment is wholly a reaction to the actions taken in the previous moment. Before doing anything, we should think carefully about all impending influences and allow ourselves to settle on the balance of those influences. This approach will allow us to optimize divergence and maintain a solid thread of causality. It is also pertinent that letting our activities be driven by causality allows us to exploit the support of existing logical evidence.

When an unpredictable, random or traumatic event occurs, causality is broken. Consequently, such an event creates the opportunity for completely free actions because there is no rational connectivity between current and prior actions. These actions represent a discontinuous causal nexus with new emergent meaning. Error is an opportunity for growth and the expansion of consciousness.

Unconscious Concentration

The goal of concentration is an extremely focused thought sequence. That focus is naturally achieved when the thought sequence or concept is more substantial and has ties to a physical representation. Therefore, it is consistent to propose that the result of concentration is an instinctual concept and this is a form of down-shifting or intrapolation of the original concept. In other words, we want the result of concentration to be a completely understood concept.

However, we find that if concentration is a product of the unconscious rather than the conscious mind, a stable, focused concept may be achieved although it remains at a relatively high frequency. Thus, the understanding of the concept is achieved within the unconscious regime so that we are never consciously aware of it until some associated insight or phenomenon occurs.

Unconscious concentration can attend to problems posed by the conscious mind and provide solutions. Those solutions become foreground thought, albeit at high frequencies and consequently, probability waves can be produced. Manifestations resulting from unconscious concentration are essentially the result of non-specific thought sequences. In other words, the original concept which ultimately generates the manifestation may be in the form of a general desire. For, example, the desire for more money with no specification of process may be unconsciously resolved into a particular solution. That solution in turn produces a manifestation of more money.

In general, the intensity of unconscious concentration is proportional to the uncertainty introduced by the original problem. To a large extent, the impetus of the mind is the reduction of uncertainty. However, unless the mind has internal connectivity, uncertainties will remain compartmentalized. Therefore, the effectiveness of unconscious concentration is promoted by holistic thinking.

Clearly, the objective of concentration, whether it is conscious or unconscious, is to develop focus. That is, we want to produce precise thinking. A mind capable of precise thinking will automatically manifest because probability distortions will be minimally divergent. The problem is that thinking is generally fuzzy and precision only comes with the proper training. However, exercises in concentration will train the mind to become more precise. We point out that there is a limit to the precision available to the mind imposed by quantum mechanics. The uncertainty principle in quantum mechanics states that there is a minimum bound to uncertainty below which we can never go. Consequently, there is a limit to the level of manifestation of even the most precise mind.

The unconscious mind categorizes every experience. That is, every perception, sensation, and concept is stored somewhere in the unconscious mind. As time progresses and more experiences are accumulated, the mind finds ways to compress information so that it becomes encoded. It is through concentration that the mind is able to examine and decode these experiences. We also find different interpretations of our experiences in our dreams. When the mind is incapable of physical awareness, it can interact with normally unconscious thought sequences which are dreams. These are high frequency thought sequences and it follows that time appears expanded in a dream.

Unconscious concentration is much more intensive than conscious concentration because the unconscious mind typically carries ninety percent of the mind's mental capacity. Consequently, manifestations produced by unconscious concentration are more powerful than typical manifestations.

The drawback of an unconscious manifestation is that we never develop a complete understanding of the process. While difficult situations may require unconscious manifestation, we encourage conscious manifestation in the interest of cultivating knowledge.

Stability

Stability is an important feature of our behavior. However, in a reality consisting of various and sundry manifestations, we are subject, either directly or indirectly to unexpected probability distortions. These distortions can create undesirable influences that irrevocably alter our paths.

We find that there are two extremes of behavior that are stable. Namely the two extremes are complex, unpredictable behavior and simple, predictable behavior. Complex, unpredictable behavior has low probabilities and consequently it can only be affected by large probability distortions. Simple, predictable behavior is essentially deterministic and the probabilities are saturated at unity. Consequently, probability distortions will have no affect on simple, predictable behavior.

Problems with stability occur for fuzzy, indecisive behavior which contain moderate probabilities. Such behavior is more subject to probability distortions.

Obviously, we are proposing that decisive or precise behavior, whether it is complex or simple is stable behavior. It is appropriate that precise thinking is consistent with both stability and successful manifestations.

Intelligence and/or education can enhance precise thinking and thus, stability. The advantages of a good education are multiple and include the ability to lead a productive and stable life.

Another advantage of stability is the ability to recover from undesirable circumstances. Such circumstances remain outliers for the stable individual rather than permanent alterations. In physics, a stable system is subject to a restoring force. That is, if the pattern of activity of a system is interrupted by an outside disturbance, then there is an intrinsic restoring force which acts to recover the pattern. In contrast, unstable systems have no substantial restoring force and their activity becomes chaotic once disturbed. This suggests that there is connection between the concept of a restoring force and precise activity.

With respect to a machine, precision implies deterministic activity. However, with respect to a human being, precision can be consistent with free will. The activity of a human being can be locally unpredictable, but globally deterministic. This means that the cycle of the activities has a long period. In that case, the activities cannot be perceived as deterministic, although they are on a larger scale.

Consequently, we are saying that the occasionally unpredictable behavior of a human being may be actually cyclic. The period of that activity may be longer than the lifetime of the individual. This proposition allows the activity to be precise, while appearing random. In comparison, it turns out that the orbits of certain asteroids are apparently erratic, but are actually cyclic and thus, stable.

This proposition is similar to the philosophical concept of eternal recurrence which purports that the universe has an infinite succession of repeated occurrences. It is also a resolution to the paradox of destiny and free will. That is, our activities are destined to the extent that they are components of a larger cycle of activities. However, because our perceivable actions are unpredictable, we experience free will.

In contrast, the behavior of some individuals may in fact be unpredictable globally due to an insufficient restoring force and the effects of arbitrary probability distortions.

Essentially, our life may consist of order or chaos. If it consists of order, then it is stable. If it consists of chaos, then it is unstable.

Probability Waves

The manifestation of an event occurs when the probability of that event is driven near unity and a subsequent energy disturbance stimulates the event into existence. The essential challenge of a manifestation is the distortion of probability. Probability is distorted by creating potentiality. If we begin a manifestation with a concept, then we must create a design or blueprint with sufficient potentiality. That is, the blueprint must have enough resolution and logical consistency for a real representation of it to occur.

We might think of a manifestation as an invention. The energy of the manifestation is a measure of the concentration needed to invent it. The specific details of the invention must be precise because even subtle irregularities in the blueprint of the invention can disable its potentiality. We refer to the precision of the invention as a property of tuning.

A requirement of the successful propagation of a manifestation is that small changes in energy can cause significant changes in probability. The way in which a probability wave propagates is as follows. The blueprint represents the original probability distortion. The probability distortion gets transmitted as a small energy disturbance. The received energy disturbance is converted into the original probability distortion.

The concentration needed to invent is a measure of the probability distortion created by the blueprint. The size of the probability distortion is proportional to the energy disturbance it creates. Therefore, the concentration is proportional to the energy disturbance. The more concentration used, the greater the energetic stimulation and thus, the more likely the manifestation.

When we are able to reflect incoming probability distortions, our thoughts are deeply encoded. Encoding allows us to detect probability distortions and objectify their potential effects on our system. When our thoughts are encoded, the mapping of identities into meanings remains private. Consequently, probability distortions are unable to alter our proclivities.

In contrast to concentration, meditation disengages outgoing probability distortions because it diffuses them. Essentially, meditation minimizes the difference between background and foreground thought which would normally create a probability distortion.

Due to the rate at which the mind multiplexes thoughts, a probability distortion oscillates. That is, the mind cycles its attention between background and foreground thought. This oscillation causes far-field spherical probability waves. The probability waves carry energy but that energy is diminished as the probability waves diverge.

There is a significant difference to the way in which information travels and probability distortions travel. Information doesn't diverge. Rather, the energy in which information is encoded either diverges or is attenuated. In contrast, probability distortions diverge. Another difference is that information is not typically radiated. It is carried by energy. Probability distortions radiate as the squared amplitudes of probability waves generated by oscillations in probability.

Without concentration to cumulatively augment the probability distortions, there would be no significant energy signatures at remote sites. It is also clear that a site needs the capability to decode the energy signatures of probability distortions.

Power

The power of a source is the rate at which a manifestation occurs.

Thus, the power of a source, P is the energy of the cumulative probability distortions, (7), divided by the time required for the manifestation, (10).

$$P = \frac{\Delta I}{\Delta t}.$$

(26)

The more a person is able to convert her thoughts into reality, the more control she has over reality. That control is a measure of her power.

In physics, power is the rate of change of energy. Thus, consistency demands that manifestation is equivalent to energy. Intelligence is equivalent to information rate, which is equivalent to energy. Consequently, intelligence is equivalent to manifestation. We can conclude that power is the rate of expression of intelligence. When intelligence remains inert, there is no power.

Active intelligence is promoted by the will. The will to power is a concept originally conceived by Nietzsche. It suggests that the ambition and drive needed to succeed in the world is created by the will. If it is true that with will, intelligence is expressed, then the underlying motive for the great scientific discoveries and the great works of art and music is the will.

Power is consistent with manipulation. Someone with a high level of skill at manipulating tools has power. A tool can be of various and sundry forms. It can be a machine or it can be an instrument or it can be a person. Machiavelli makes the use of political power as a form of manipulation famous in his book *The Prince.*

Although power is often characterized as evil, it has no particular moral disposition. It can be used for self-help or to help others or to hurt others. Power in the right hands is considered to be a great good.

Accordingly, while manifestations are generally used for self-help, they can be very beneficial for others. For example, a person becomes rich through manifestation and subsequently creates jobs. As another example, consider a scientist who manifests a solution to a major problem. That solution then provides resources for others.

Bacon coined the phrase "knowledge is power." However, we find that knowledge is equivalent to energy since it is information gathered over time. Rather, it is the rate of expression of knowledge that is power. That is, power is the rate at which knowledge is used. Consequently, when knowledge is unused, it has no power. We point out that the difference between intelligence and knowledge is that intelligence is the instantaneous rate of change of information while knowledge is the average rate of change of information. Knowledge requires a history of information while intelligence doesn't. If information is denoted by i, and knowledge is denoted by K, then we can write

$$I = \frac{di}{dt}$$

(27)

and

$$K = \frac{\Delta i}{\Delta t}.$$

(28)

It follows that instantaneous power is the derivative of intelligence.

$$P = \frac{d^2 i}{dt^2}.$$

(29)

We might say that power is the rate at which intelligence is converted from potential energy into kinetic energy. This is consistent with the rate of a manifestation (26). The total energy remains a constant.

Interference

Interference occurs when actions interact either to augment the net action or to diminish the net action. When the net action is augmented, there is constructive interference. When the net action is diminished, there is destructive interference. In physics, interference effects occur with waves, however, in general, interference may apply to ordinary actions.

There are two ways for probability waves to interfere with each other. The waves from different sources may directly interfere or different probability distortions may interfere through the principle of complimentarity.

A wave will have a different amplitude depending on its phase, φ. Thus, referring to the probability wave given in (11), we might have perfectly constructive interference if

$$u_n(\varphi_1) + u_n(\varphi_2) = 2u_n .$$
(30)

Alternatively, it is possible that the interference is perfectly destructive

$$u_n(\varphi_1) + u_n(\varphi_2) = 0 .$$
(31)

Consider the case when a multitude of probability waves interfere at one point in space, r_0. Given that the phases are random, the interference is largely destructive. However, since the source located at r_0 can have only one phase, that particular source wave will survive. That is,

$$\sum_{n,v} u_n(\varphi_v) = u_0$$

$$\tag{32}$$

Consequently, at a given point in space, a multitude of probability waves from different sources with different phases will result in destructive interference, leaving the source wave emanating from the source located at the given point as the residual. We can surmise that when multiple sources are trying to manifest at the same site, the only successful source will be the one located at the site. This result is reinforced by the fact that there will be no divergence for this particular source.

Alternatively, we find that when there are multiple sources, but there is no source at the site of interference, the net amplitude will be zero. That is,

$$\sum_{n \neq 0,v} u_n(\varphi_v) = 0$$

$$\tag{33}$$

Even if the probability waves do not interfere, the probability distortions they produce may interfere. Consider the case of a binary system that can have only two states. The probability of state one is p_1 and the probability of state two is p_2. We must have

$$p_1 + p_2 = 1$$

$$\tag{34}$$

A probability distortion tuned to state one must distribute itself such that

$$(p_1 + \delta p) + (p_2 - \delta p) = 1$$

$$\tag{35}$$

A subsequent probability distortion tuned to state two must distribute itself such that

$$(p_1 + \delta p) - \delta p + (p_2 - \delta p) + \delta p = 1 \tag{36}$$

which is the original configuration given by (34). Equation (36) demonstrates a case of destructive interference due to complimentary probability distortions. Thus, we find that motivations to do complimentary actions will be equivalent to no net motivation.

Tuning

Given a source that transmits probability waves to a remote site, the probability distortions are directed to specific probabilities within the site's probability distribution by tuning. Essentially, tuning is a phase adjustment.

$$u_n \rightarrow e^{i\delta\varphi} u_n .$$

(37)

That is, the source must adjust its phase with a correction $\delta\varphi$.

Tuning is the correlation of probability distortions transmitted by the source to the probability distortions decoded at the site.

Recall that the transmitted distortions are absorbed as energy and then decoded back into probability distortions. This is similar to the way in which the communication process works. That is, a mind transforms thoughts into energy signals and transmits them to the receiving mind which decodes the signals back into thoughts.

It is clear that normally, the mind is perfectly tuned to the body. If I want to move my right foot, I think about it and my right foot moves. If my left foot moved instead of my right foot, then my mind would not be tuned to my body. It is similar for manifestations. If I wanted to manifest money for myself, but instead, I lost money, then my manifestation wasn't properly tuned. If I wanted to manifest a date with a certain woman, but instead, I got a date with her friend, then my manifestation wasn't properly tuned.

Tuning does not alter the "energy" of a probability wave, so the strength of the probability distortion is unaffected. That is,

$$\left(e^{i\delta\varphi}u_n\right)\left(e^{-i\delta\varphi}u_n^*\right)=u_n^2.$$

$$u_n^2=\delta p_n. \tag{38}$$

A Summary of the Theory

Concentration by a source creates a probability distortion given as

$$\delta p_0 = \sum_{i=1} p_i - p_0$$

.

Due to multiplexing, the distortion generates a spherical wave given as

$$u_n = r_0 \frac{\sqrt{\delta p_0}}{r_n} e^{i\left(\omega t - \vec{k}.\vec{r}_n\right)}$$

.

This wave is properly tuned.

$$u_n \rightarrow e^{i\delta\varphi} u_n$$

.

The probability distortion is the "energy" of the wave

$$\delta p_n = r_0^2 \frac{\delta p_0}{r_n^2}$$

The cumulative distortions have an equivalent true energy given as

$$\Delta I = -\hbar\omega \sum_n \log\left(1 - \delta p_n\right)$$

.

This energy is deposited at a site and decoded into the original cumulative distortions.

$$\Delta I \to \sum_n \delta p_n .$$

The cumulative probability distortions alter the probability distribution at the remote site.

$$p_{ijk}^{(l)}(t) + \sum_n \delta p_n \to p_{ijk}^{(m)}(t + \Delta t) .$$

The rate of manifestation is the power of the source:

$$P = \frac{\Delta I}{\Delta t} .$$

Made in the USA
Columbia, SC
21 March 2020